Advances in Environmental Engineering

Advances in Environmental Engineering

Special Issue Editors

Adriana Estokova
Magdalena Balintova

MDPI • Basel • Beijing • Wuhan • Barcelona • Belgrade

MDPI

Special Issue Editors

Adriana Estokova
Technical University of Kosice
Slovakia

Magdalena Balintova
Technical University of Kosice
Slovakia

Editorial Office
MDPI
St. Alban-Anlage 66
Basel, Switzerland

This is a reprint of articles from the Special Issue published online in the open access journal *Environments* (ISSN 2076-3298) from 2017 to 2018 (available at: http://www.mdpi.com/journal/environments/special_issues/env_eng)

For citation purposes, cite each article independently as indicated on the article page online and as indicated below:

LastName, A.A.; LastName, B.B.; LastName, C.C. Article Title. *Journal Name* **Year**, *Article Number*, Page Range.

ISBN 978-3-03897-001-9 (Pbk)
ISBN 978-3-03897-002-6 (PDF)

Cover image courtesy of Adriana Estokova and Magdalena Balintova.

Contents

About the Special Issue Editors

Adriana Estokova, Prof. Dr., graduated in Inorganic Chemistry and finished her Ph.D. studies in Environmental Engineering. She became an Associate Professor and a Professor of Environmental Engineering. Currently, she is a full Professor at the Faculty of Civil Engineering at the Technical University of Kosice, Slovakia and she is a Head of the Department of Material Engineering. She published more than 300 publications, 91 of them are included in the Scopus database and 78 in the Web of Science database. Her papers have received more than 240 citations. She is a member of several associations and journals, editorial boards as well as of scientific committees of international and national conferences. She serves as an editor, guest editor and reviewer for several prestigious scientific journals. Her professional experience is in the field of environmental and material engineering. Her research is focused on the assessment of the environmental safety of building materials (heavy metals, leachability, and radionuclide activity) and the environmental impacts of buildings (LCA, indoor air), as well as on the durability of cement composites in an aggressive environment (bio-corrosion and chemical corrosion).

Magdalena Balinova, Prof. Dr., graduated in inorganic chemistry and finished her Ph.D. studies in Mineralurgy and Environmental Technologies. She became an Associate Professor and a Professor in Environmental Engineering. Currently, she is a full Professor at the Faculty of Civil Engineering at Technical University of Kosice, Slovakia and she is a head of the Institute of Environmental Engineering. She has published more than 300 publications, 78 of them are included in Scopus database and 67 in Web of Science database. Her papers have received more than 260 citations. She is a member of several associations and journals, editorial boards as well as of scientific committees of international and national conferences. She serves as an editor, guest editor and reviewer for several prestigious scientific journals. Her professional experience is in the field of environmental engineering. The research is focused on waste water treatment, water and sediment quality, remediation methods, heavy metals removal, and soil protection.

Preface to "Advances in Environmental Engineering"

The contamination of territory by harmful pollutants coming from human activities poses a significant risk to human health and to major components of the environment. The role of environmental engineering is to apply scientific and engineering principles to minimize the adverse effects of human activity on the environment and to improve the quality of a natural environment to provide sustainable water, air, and land quality for human habitation and for other organisms. At present, environmental engineering is mainly focused on environmental technologies for cleaner production, allowing the reduction of emissions and waste production, the usage of renewable resources, recycling waste, Mitigation of the effects of natural disasters (floods, droughts), sustainable urban planning and development, and others.

The strong interest in protecting the environment has created new responsibilities for scientists from a wide range of research fields. This book presents research papers providing an integrated view of the trends in solving the problems associated with the achievement of sustainability in environmental engineering. Special attention is paid to advanced waste water technologies, soil protection, and sediment pollution modelling, environmental impacts of technologies, life cycle analysis (LCA), air quality and indoor environment, and special applications of environmental materials.

Adriana Estokova, Magdalena Balintova
Special Issue Editors

environments

MDPI

Editorial

Advances in Environmental Engineering

Adriana Estokova * and Magdalena Balintova

Faculty of Civil Engineering, Institute of Environmental Engineering, Technical University of Kosice, Vysokoskolska 4, 04200 Kosice, Slovakia; magdalena.balintova@tuke.sk
* Correspondence: adriana.estokova@tuke.sk; Tel.: +421-55-602-4265

Received: 25 April 2018; Accepted: 30 April 2018; Published: 4 May 2018

Environmental quality is crucial to our health, our economy, and our lives. However, it faces several serious challenges, not least those of climate change, unsustainable consumption, and production, as well as various forms of pollution. This Special Issue collects research papers aimed at a wide range of environmental topics: water and wastewater treatment and management, soil degradation and conservation, sediment pollution control, the environmental impact of technologies, life cycle analysis (LCA), air quality and indoor environment, and advanced environmental materials. Contributions describe novel and significant knowledge, scientific results, and advanced applications in the field of environmental engineering. This Special Issue provides an integrated view of the trends in solving the problems associated with the achievement of sustainability in environmental engineering. This issue contains twelve papers that have been selected as emerging studies dealing with the above-mentioned topics.

The contributions, aimed at wastewater treatment, present a wide range of methods applied to various pollution removal methods. Investigating contaminants of emerging concern such as pharmaceuticals and personal care products reveals that the fate of these compounds in the aquatic environment has been a topic of wide interest and active research. Lecours et al. [1] applied different electrochemical approaches to the study of the oxidation products of the anti-infective trimethoprim, a contaminant of emerging concern frequently reported in wastewaters and surface waters. The authors found that electrochemical techniques are relevant not only to mimicking specific biotransformation reactions of organic contaminants but also to studying the oxidation reactions of organic contaminants of interest in water treatment.

Tian et al. [2] investigated the effects of physico-chemical post-treatments of sewage sludge using ultrasonic, ultrasonic-ozone, and ultrasonic+alkaline methods. The results showed that the post-treatments were able to increase biogas production and decrease the amount of volatile solids in the final effluent.

In the work by Pipiska et al. [3], the biosorption methods for pollution removal from wastewater were studied. Dried biomass of freshwater moss *V. dubyana* has been used as biosorbent for cationic dyes methylene blue and thioflavin T removal from both single and binary systems. Influence of a contact time, pH, and sorbate concentration on the dyes' removal efficiency has been investigated. The authors reported that an increase in pH has a positive effect on both thioflavin T and methylene blue sorption, and dye removal by moss *V. dubyana* is likely based on the electrostatic attraction.

Another paper [4], dealing with wastewater treatment, studied and quantified the elimination of sunflower oil from wastewater influent using a biological treatment involving activated sludge. The findings revealed that the efficiency of the elimination of sunflower oil using a combination of biodegradation and flotation was 90%.

The next two papers deal with soil properties and sediment modelling. Gomboš et al. studied the selected parameters of soils for further numerical simulation of the water regime and its prognosis under heavy soil conditions. Soil profiles were evaluated for the distribution of volume changes to the horizontal and vertical components. The effect of texture on geometric factor values was analysed.

A close correlation between the measured horizontal volume changes and the geometric factor value has been found [5].

Junakova et al. focused on the design of the mathematical model that was intended to predict the total content of nitrogen, phosphorus, and potassium in bottom sediments in small water reservoirs depending on water erosion processes. The proposed model was validated in the small agricultural watershed of the Tisovec River, Slovakia. The results indicate the applicability of the new model in predicting the quality of the reservoir's sediment detached through erosion processes in the watershed [6].

The environmental impact of various technologies has been assessed in the next three papers. The paper [7] deals with the life cycle assessment of electricity generation from various energy sources in the Czech Republic. The greenhouse gas emissions were chosen as key indicators to evaluate the environmental load of particular energy sources. The obtained results revealed that the worst environmental impact in terms of greenhouse gas emissions is linked to electricity generation based on lignite.

Zeleňáková et al. [8] reported on the environmental impact of a small hydro power plant including the selection of the optimal alternative of the assessed construction and proposed measurements to reduce the negative impact. Their paper points to the importance of assessing the impact of construction on the environment in the early planning phase. Eliminating the negative environmental impact of construction in the early phase of design is much more challenging than the implementation of measures in the construction or operation phases.

The variant solutions of a selected heating system were analysed by Ondrejka Harbulakova et al. [9] using methods of the environmental impact assessment (EIA). Multi-criteria analysis proved that the construction of the biomass-fired power plant was the most suitable solution among three assessed variants (zero alternative, biomass power plant, and modernized gas boiler).

A significant negative impact on human health and the quality of life of millions of people worldwide is associated with urban air pollution. Tsai [10] presents an overview of the Taiwan's air quality with a special regard to the indoor air. The paper points to the importance of using green building materials in terms of the low emission of volatile organic compounds (VOCs) and other air toxics occurring indoors. The author highlights Taiwan's efforts to indoor air quality improvement through legal systems and promotion measures, which are relevant to the contribution to the quality and sustainability of the environment.

Other dangerous pollutants in the air are particulate matters of various origins. Road traffic emissions caused by both exhaust and non-exhaust sources contribute significantly to the particulate matter (PM) concentration in an urban atmosphere. Penkała et al. [11] reported that direct road-surface abrasion is of minor importance when the road is undamaged. However, the paper analyses the impact of abrasion emission reflecting realistic conditions, analysing how such emission changes, both quantitatively and qualitatively, the character of PM near roads.

With the development of new urban areas, there is a great challenge in finding new materials with an environmental added value. Pervious concrete can be an environmental solution for managing storm-water runoff. Kovac et al. [12] presented an alternative method for storm-water control using porous pavements. This paper presents the results of experimental work aimed at testing technically important properties of pervious concrete prepared with three different water-to-cement ratios. The results show that a decrease in water-to-cement ratio caused only slight differences in strength characteristics.

Author Contributions: A.E. processed an overview of the papers with topics focused on air quality and environmental evaluation of the technologies. M.B. processed an overview of the papers with topics focused on water and soil quality and wastewater treatment.

Acknowledgments: The editors would like to thank all the authors participated on the Special Issue.

Conflicts of Interest: The authors declare no conflict of interest.

References

1. Lecours, M.-A.; Eysseric, E.; Yargeau, V.; Lessard, J.; Brisard, G.M.; Segura, P.A. Electrochemistry-High Resolution Mass Spectrometry to Study Oxidation Products of Trimethoprim. *Environments* **2018**, *5*, 18. [CrossRef]
2. Tian, X.; Trzcinski, A. Effects of Physico-Chemical Post-Treatments on the Semi-Continuous Anaerobic Digestion of Sewage Sludge. *Environments* **2017**, *4*, 49. [CrossRef]
3. Pipíška, M.; Valica, M.; Partelová, D.; Horník, M.; Lesný, J.; Hostin, S. Removal of Synthetic Dyes by Dried Biomass of Freshwater Moss *Vesicularia Dubyana*: A Batch Biosorption Study. *Environments* **2018**, *5*, 10. [CrossRef]
4. Cisterna, P. Biological Treatment by Active Sludge with High Biomass Concentration at Laboratory Scale for Mixed Inflow of Sunflower Oil and Saccharose. *Environments* **2017**, *4*, 69. [CrossRef]
5. Gomboš, M.; Tall, A.; Kandra, B.; Balejčíková, L.; Pavelková, D. Geometric Factor as the Characteristics of the Three-Dimensional Process of Volume Changes of Heavy Soils. *Environments* **2018**, *5*, 45. [CrossRef]
6. Junakova, N.; Balintova, M.; Vodička, R.; Junak, J. Prediction of Reservoir Sediment Quality Based on Erosion Processes in Watershed Using Mathematical Modelling. *Environments* **2018**, *5*, 6. [CrossRef]
7. Jursová, S.; Burchart-Korol, D.; Pustějovská, P.; Korol, J.; Blaut, A. Greenhouse Gas Emission Assessment from Electricity Production in the Czech Republic. *Environments* **2018**, *5*, 17. [CrossRef]
8. Zeleňáková, M.; Fijko, R.; Diaconu, D.C.; Remeňáková, I. Environmental Impact of Small Hydro Power Plant—A Case Study. *Environments* **2018**, *5*, 12. [CrossRef]
9. Ondrejka Harbulakova, V.; Zelenakova, M.; Purcz, P.; Olejnik, A. Selection of the Best Alternative of Heating System by Environmental Impact Assessment—Case Study. *Environments* **2018**, *5*, 19. [CrossRef]
10. Tsai, W.-T. Overview of Green Building Material (GBM) Policies and Guidelines with Relevance to Indoor Air Quality Management in Taiwan. *Environments* **2018**, *5*, 4. [CrossRef]
11. Penkała, M.; Ogrodnik, P.; Rogula-Kozłowska, W. Particulate Matter from the Road Surface Abrasion as a Problem of Non-Exhaust Emission Control. *Environments* **2018**, *5*, 9. [CrossRef]
12. Kováč, M.; Sičáková, A. Pervious Concrete as an Environmental Solution for Pavements: Focus on Key Properties. *Environments* **2018**, *5*, 11. [CrossRef]

environments

MDPI

Article

Electrochemistry-High Resolution Mass Spectrometry to Study Oxidation Products of Trimethoprim

Marc-André Lecours [1], Emmanuel Eysseric [1], Viviane Yargeau [2], Jean Lessard [1], Gessie M. Brisard [1] (iD) and Pedro A. Segura [1,*] (iD)

[1] Department of Chemistry, Université de Sherbrooke, Sherbrooke, QC J1K 2R1, Canada; marc-andre.lecours@usherbrooke.ca (M.-A.L.); Emmanuel.Eysseric@USherbrooke.ca (E.E.); Jean.Lessard@USherbrooke.ca (J.L.); Gessie.Brisard@USherbrooke.ca (G.M.B.)
[2] Department of Chemical Engineering, McGill University, Montreal, QC H3A 2B2, Canada; viviane.yargeau@mcgill.ca
* Correspondence: pa.segura@usherbrooke.ca; Tel.: +1-819-821-7922; Fax: +1-819-821-8019

Received: 26 November 2017; Accepted: 19 January 2018; Published: 24 January 2018

Abstract: The study of the fate of emerging organic contaminants (EOCs), especially the identification of transformation products, after water treatment or in the aquatic environment, is a topic of growing interest. In recent years, electrochemistry coupled to mass spectrometry has attracted a lot of attention as an alternative technique to investigate oxidation metabolites of organic compounds. The present study used different electrochemical approaches, such as cyclic voltammetry, electrolysis, electro-assisted Fenton reaction coupled offline to high resolution mass spectrometry and thin-layer flow cell coupled online to high resolution mass spectrometry, to study oxidation products of the anti-infective trimethoprim, a contaminant of emerging concern frequently reported in wastewaters and surface waters. Results showed that mono- and di-hydroxylated derivatives of trimethoprim were generated in electrochemically and possibly tri-hydroxylated derivatives as well. Those compounds have been previously reported as mammalian and bacterial metabolites as well as transformation products of advance oxidation processes applied to waters containing trimethoprim. Therefore, this study confirmed that electrochemical techniques are relevant not only to mimic specific biotransformation reactions of organic contaminants, as it has been suggested previously, but also to study the oxidation reactions of organic contaminants of interest in water treatment. The key role that redox reactions play in the environment make electrochemistry-high resolution mass spectrometry a sensitive and simple technique to improve our understanding of the fate of organic contaminants in the environment.

Keywords: redox reactions; EC-MS; Fenton reaction; fate of contaminants of emerging concern; transformation products

---

1. Introduction

In the late 1990s, researchers started to demonstrate the importance of investigating contaminants of emerging concern (EOCs) such as pharmaceuticals and personal care products and since then, the fate of these compounds in wastewater treatment plants and in the aquatic environment has been a topic of wide interest and active research. It is known that enzymes, such as those of the cytochrome P450 (CYP) super family, catalyze many oxidative reactions that transform EOCs during secondary (biological) water treatment processes or after these compounds are released into the natural environment [1]. The conventional method of the study of biotransformation products of organic contaminants involves extracting transformation products from in vivo experiments, or performing in vitro experiments using cellular extracts containing CYP450 enzymes [2,3]. In order to remove uninteresting compounds which might interfere with the analysis, reduce sample complexity and simplify interpretation of the results, those approaches require laborious sample preparation to detect the compounds of interest [4]. In addition, performing experiments with biological systems requires the use of low concentrations of contaminants to avoid toxic effects leading to low concentrations of transformation products, making their identification and characterization even more difficult. Besides biological transformations, EOCs may be transformed after tertiary processes during wastewater treatment such as ozonation, ultraviolet light or advanced oxidation processes, i.e., chemical oxidation processes occurring via reactions with hydroxyl radicals [5]. Considering that these approaches do not generally result in complete mineralization under usual treatment conditions, these processes lead to the formation of unknown compounds [6,7]. To improve risk assessments for aquatic biota, it is important to understand the mechanisms of formation of those oxidation products as well as to elucidate their molecular structure [8].

Electrochemistry coupled offline or online with mass spectrometry is an interesting alternative to study oxidation products of EOCs, since experiments are done in controlled conditions using pure solvents and reagents and higher concentration of contaminants, reducing or eliminating the need for sample preparation and accelerating data analysis and identification workflows [9]. Mass spectrometry is well suited for coupling with electrochemistry given the rapidity, sensitivity and specificity of modern mass spectrometers. Studies have demonstrated that metabolites generated by enzyme-catalyzed reactions such as *N*-dealkylation, *N*-oxidation and *O*-dealkylation, aliphatic hydroxylation and aromatic hydroxylation can be generated in vitro by electrochemical methods [9,10]. The similarity between results obtained by such apparently unrelated systems (enzymatic vs. electrochemical cell) is explained by the underlying mechanisms that occur in the phase I metabolism, which are generally initiated by single electron transfer or hydrogen atom transfer involving an iron-oxygen complex [9,11]. For example, in electrochemical cells, single-electron transfer mechanisms such as the one occurring during *N*-dealkylation can be reproduced by oxidation reactions at the working electrode in the presence of a basic supporting electrolyte [11]. Electrochemistry was showed also to be useful to improve our understanding of abiotic processes that degrade or cause EOCs in the environment to bind to soils, as demonstrated by Hoffmann, et al. [12] with the sulfonamide antimicrobial sulfadiazine. It is clear at this point that electrochemistry cannot mimic all possible CYP450-catalyzed reactions, in fact only those reactions initiated by single-electron transfer (N-, O-, and S-dealkylation, hydroxylation of benzylic carbon, etc.) can be simulated by electrochemistry [13]. However, electrochemistry can be useful as a starting point to tackle the complexity found in samples issued from natural transformation processes, as it was demonstrated in a study on the transformation products of an EOC produced by the White-Rot Fungus *Pleurotus ostreatus* [14]. In that study, the formation of multiple complex biotransformation products of carbamazepine by electrochemistry, such as epoxy, dihydro and methoxy derivatives, was demonstrated using an online electrochemistry–mass spectrometry technique. According to the authors of that paper, workflows of identification of transformation products of organic contaminants can be improved by electrochemistry coupled to mass spectrometry.

The objective of the present work was to study the oxidation of a common EOC while demonstrating the usefulness of using electrochemistry coupled offline or online with mass spectrometry to improve our understanding of the fate of organic contaminants in the environment. Three different electrochemical experiments coupled offline to high-resolution mass spectrometry (HRMS) were investigated: cyclic voltammetry, electrolysis and the electro-assisted Fenton reaction. Also, an electrochemical experiment using a thin-layer flow cell coupled online to HRMS was performed. The anti-infective trimethoprim (TRI) was chosen as model compound given its frequent detection in environmental waters [15] and the availability in the literature of information on its oxidation products [16–21] and metabolites [22,23].

2. Material and Methods

2.1. Reagents and Chemicals

Trimethoprim was purchased from Santa-Cruz Biotechnology (Dallas, TX, USA). Tetrabutylammonium perchlorate (TBAP) was acquired from TCI America (Portland, OR, USA). 2,4-Diaminopyrimidine, 3,4,5-trimethoxytoluene, iron (II) sulfate and sodium sulfate were obtained from Sigma-Aldrich (St. Louis, MO, USA). All these products have a purity ≥98%. Water for electrochemistry experiments was purified using a Milli-Q filtration system (Merck MIllipore, Burlington, MA, USA). Solvents and additives used in liquid chromatography–mass spectrometry experiments, such as acetonitrile (ACN), water and formic acid (FA), were purchased from Fisher Scientific (Ottawa, ON, Canada) and were Optima LC/MS grade.

2.2. Cyclic Voltammetry and Electrolysis

A three-compartment glass cell, each containing its own electrode, was used for both cyclic voltammetry (CV) and electrolysis experiments. The counter electrode, a Pt mesh connected to a Pt wire, was separated from the working electrode by a fine-sized glass frit. The reference electrode communicates with the central working compartment (volume ≈ 25 mL) via a Luggin capillary placed just below the surface of the working electrode. Working electrodes for cyclic voltammetry experiments had a 2 mm diameter and were made of glassy carbon, Au or Pt. For electrolysis experiments, the Au electrode was 6 mm and the glassy carbon 7 mm in diameter to improve the rate of transformation of TRI. The Ag/Ag^+ reference electrode was prepared fresh on each day of the experiments. The reference solution consisted of 1 mL of ACN:H_2O 99:1 (v/v) solution with 100 mM (34.2 g·L^{-1}) TBAP as the supporting electrolyte to which 1 mM (169.9 mg·L^{-1}) $AgNO_3$ was added. An Ag wire was rubbed with fine sandpaper to remove the oxide layer and expose a fresh Ag surface. The wire was rinsed with ACN: H_2O 99:1 (v/v) before being immersed in the reference solution to form the reversible Ag/Ag^+ redox couple. Organic medium such as ACN and TBAP is usually chosen to have access to a large potential window (very negative and very positive potential), this combination of solvent and salt is very stable in a wide range of potentials and various electrode materials. The presence of water (1% here) is added to furnish little and controlled amounts of protons.

To study the electrochemical behavior of TRI, CV experiments were used first to determine the general redox pattern of TRI and to monitor the potentials to be applied during subsequent electrolysis (constant potential) experiments. All experiments were performed at room temperature. Cyclic voltammetry experiments were carried out in a purely diffusional regime (working electrode was stationary) using conditions used in previous electrochemical oxidation studies [24,25]. Solutions of TRI (1 mM, 290.3 mg·L^{-1}) were prepared in ACN:H_2O 99:1 (v/v) with 100 mM TBAP. The supporting electrolyte was set at 100 mM to ensure a good conductivity of the solution. To eliminate reactions of radical intermediates with dissolved O_2, the working electrode compartment (anode compartment) was maintained under a $N_{2(g)}$ atmosphere. $N_{2(g)}$ was passed through a bubbler containing ACN to limit evaporation during prolonged experiments and the flow rate was adjusted to achieve moderate bubbling to limit undesirable convection phenomena. The cathode compartment was open to air. A CV of a blank solution of TBAP/ACN without TRI was always recorded before each experiment. The potential scan rate was set at 50 or 100 mV·s^{-1} for all CV experiments.

Electrolysis experiments were done at constant potential and were conducted with a potentiostat/galvanostat EG&G model 273A from Princeton Applied Research. The solution was stirred to increase the supply of electroactive compound to the surface of the working electrode and to promote a higher conversion rate. Also, a higher concentration of TRI, 10 mM (2903 mg·L^{-1}) was used. After electrolysis, samples were diluted by a factor of 1000 using water and injected into the liquid chromatography–quadrupole–time of flight mass spectrometry (LC-QqTOFMS) system (Bruker, Billerica, MA, USA). This dilution step was necessary to avoid overloading the chromatographic column with the supporting electrolyte and to prevent signal saturation for the analyte.

2.3. Electro-Assisted Fenton Reaction

Electro-assisted Fenton reaction experiments were carried out with a potentiostat/galvanostat EG&G model 273A from Princeton Applied Research in a cell with one compartment and two electrodes. The working electrode was made of glassy carbon (diameter: 7 mm) and molecular oxygen is continuously bubbled into the solution during the experiment. In the electro-assisted Fenton reaction, hydrogen peroxide is generated in situ electrochemically in an acid medium from the dissolved oxygen:

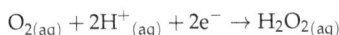

$$O_{2(aq)} + 2H^+{}_{(aq)} + 2e^- \rightarrow H_2O_{2(aq)}$$

Compared to other materials, glassy carbon shows a reduced overpotential with respect to the production of peroxide compared to other reactions, making it an ideal candidate [26] for this type of experiment. The hydroxyl radical (OH•) produced in the Fenton reaction reacts quickly with organic compounds in solution. For example, according to Dodd, et al. [27], the second order rate constant of the reaction of OH• with TRI is $(6.9 \pm 0.2) \times 10^9$ M·s^{-1} at neutral pH and at 25 °C. When produced in sufficient quantity over a sufficient period of time, OH• will oxidize virtually all carbons in the molecule to lead to an almost complete mineralization of TRI. The counter electrode/reference electrode was a Pt mesh. The solution contained 50 mM Na$_2$SO$_4$, 0.1 mM FeSO$_4$ in water acidified to pH 2.7 using concentrated H$_2$SO$_4$. Since degradation of TRI in the electro-assisted reaction is fast, a higher concentration of TRI (58 mg·L^{-1}) was used in this experiment to follow changes in its concentration throughout the duration of the whole experiment. A current density of 1 mA·cm^{-2} was applied for 30 min in galvanostatic mode. The reaction time and the current density were optimized to limit bubble formation at the counter electrode and to produce about 70% reduction of the signal of the precursor ion detected by mass spectrometry. Longer electrolysis time, e.g., >60 min, were not used to avoid further degradation of the transformation products which would then be harder to detect and characterize. All experiments were done at room temperature. Aliquots at different time intervals were sampled from the cell, diluted by a factor of 1000 and injected into the LC-QqTOFMS system.

2.4. Thin-Layer Flow Cell Coupled Online to High-Resolution Mass Spectrometry

The Roxy EC for MS System manufactured by Antec Leyden B.V. (Zoeterwoude, the Netherlands) was used to perform online EC-MS experiments. This system is composed of a syringe pump, a thin-layer flow cell (μ-PrepCell) and a three-electrode setup controlled by a potentiostat. The thin-layer flow cell has a 11 μL internal volume. The reference electrode was Pd/H$_2$, the counter electrode was Ti and the working electrode was boron-doped diamond (dimensions 12 × 30 mm, thickness 1 mm). The cell was operated in the constant potential steps mode. All experiments were done at room temperature. The outlet of the thin-layer flow cell was connected to the inlet of the electrospray source of the QqTOFMS through a PEEK tubing (127 μm internal diameter). TRI at a concentration of 4.3 μM (1.25 mg L^{-1}) dissolved in a solution of 0.1% FA in H$_2$O:ACN 1:1 (pH 2.92) was introduced in the cell using in a syringe pump at a flow rate of 20 μL min^{-1}. This concentration was optimal to avoid saturating the QqTOFMS detector. The composition of the solution used for these experiments (0.1% FA in ACN:H$_2$O 1:1 *v/v*) was chosen according to manufacturer's recommendations [28].

2.5. Liquid Chromatography–Quadrupole–Time of Flight Mass Spectrometry (LC-QqTOFMS)

Chromatographic separation of electrolysis or Fenton reaction products was performed on a Nexera ultra performance liquid chromatograph (UPLC) manufactured by Shimadzu (Kyoto, Japan) coupled to a Bruker MaXis time-of-flight mass spectrometer (QqTOFMS) equipped with an electrospray ionization (ESI) source operated in the positive mode. Reaction products generated in the thin-layer flow cell were not separated chromatographically since the cell is coupled online to the QqTOFMS and reaction products are monitored in real time.

The UPLC parameters were the following: the column was an Acquity HSS-T3 C_{18} reverse phase (50 × 2.1 mm, 1.8 μm), the mobile phase was composed of solvent A (0.1% FA in H_2O) and solvent B (0.1% FA in ACN) and the mobile phase flow rate was 500 μL·min^{-1}. The elution gradient was adjusted according to the samples to maximize separation. Therefore, two different gradients were employed for optimal separation: one for electrolysis products and the another for electro-assisted Fenton reaction products. For the analysis of electrolysis products, the gradient as percentage of B in the mobile phase was: at 0 min, 10%; 5.30 min, 15%; 7 min, 40%; 9 min, 98%; 10 min, 98%; 11 min, 10%; 14 min, 10%. For the analysis of Fenton reaction products, the gradient used was the following: 0 min, 5%; 5 min, 20%; 7 min, 50%; 8 min, 98%; 10 min, 98%; 11 min, 5%; 14 min, 98%. For both chromatographic methods the column temperature was set to 30 °C.

The injection volume was also adjusted according to the experiment and the sample. Thus between 0.1 and 2 μL of the diluted sample was injected to obtain a signal with a target intensity of 2×10^5 for the most abundant species. A switching valve was used to bypass supporting electrolyte (TBAP) which remains sufficiently concentrated to saturate the QqTOFMS detector.

The QqTOFMS parameters were the following: nebulizing gas N_2, nebulizing gas temperature 200 °C, nebulizing gas flow rate 9 L·min^{-1}, capillary voltage 3500 V, end plate offset -500 V, ion cooler 35 μs and RF 55 Vpp. The mass range was m/z 100 to 700. For the MS/MS experiments, a time segment comprising the peak of the analyte was created and the isolation window of the precursor ion was set to 1 Da. The collision energies (between 10 and 30 eV) were optimized to obtain 10% relative intensity of the precursor ion. The mass resolution measured at full width at half-maximum for m/z 291 was about 18,000.

2.6. Identification of Oxidation Products

To identify the oxidation products generated in the diverse electrochemical setups. a workflow based on the confidence level scheme proposed by Schymanski et al. [29] was adopted. According to that scheme, accurate mass represents the lowest confidence (level 5), followed by unequivocal molecular formula (level 4), tentative candidate (level 3, based on complementary data such as tandem mass spectra and software tools), probable structure (level 2a, reached using a library spectrum match or level 2b reached using experiments indicating that no other structure fits the data) and finally confirmed structure (level 1, which requires a reference standard). Since tandem mass (MS/MS) spectra of oxidation products of trimethoprim are not yet stored in spectral libraries and standards are rare or not commercially available, diverse techniques were used to improve the identification level of the accurate mass data obtained by the high-resolution mass spectrometer. Those complementary techniques were: hydrogen–deuterium exchange (HDX) [30], which reveals the number of exchangeable hydrogen atoms such as those present in alcohol or amine functional groups; MS/MS experiments and comparison with MS/MS data found in the literature; in-silico fragmentation analysis of precursor ions based on the Mass Frontier software from HighChem and spectral accuracy determined by MassWorks software from Cerno Bioscience. Spectral accuracy measures the similarity between mathematically transformed experimental isotopic pattern of an ion and the theoretical isotopic pattern corresponding to a given molecular formula [31]. High spectral accuracy, generally $\geq 98\%$, means that the experimental isotopic pattern closely matches the abundance and shape of the isotopic pattern expected for a possible formula. Therefore, spectral accuracy contributes to eliminate possible candidates and gives higher confidence in the assignation of unique molecular formulas to accurate masses. Parameters for spectral accuracy were the following: elements (C, H, N and O); number of elements determined by empirical rules, mass tolerance (10 Da); charge (+1) and number of double bond equivalents (3.5 to 11.5).

Parameters for the in-silico fragmentation analysis using Mass Frontier software were the following: ionization method ($M + H^+$); ionization on non-bonding electrons and π-bonds; cleavage (α and inductive); H-rearrangement (charge remote rearrangement, hydrogen transfer from atom α, β and γ); resonance reaction (electron sharing, charge stabilization); aromatic system allowed (ionization, stabilization); cleavage allowed on primary, secondary and tertiary carbocation; maximum reaction steps (5) and reactions limit (10).

3. Results and Discussion

3.1. Cyclic Voltammetry and Electrolysis

The results of CV measurements with TRI are shown in Figure 1. They revealed two main peaks in the voltammogram at 900 and 1150 mV (vs. Ag/Ag+). Those peaks indicate the oxidation of two electroactive functions on the TRI molecule and the formation of two oxidation products at the electrode surface. Electrolysis experiments based on those two potentials were performed to produce a sufficient quantity of oxidation products for identification by offline LC-QqTOFMS. Minor peaks were also observed at negative potentials which may have been the result of an incomplete purge of O_2 by N_2 bubbling. Nevertheless, the presence of possible traces of O_2 did not influence TRI oxidation.

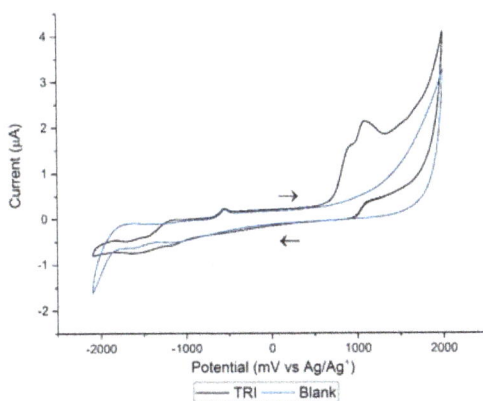

Figure 1. Cyclic voltammogram of trimethoprim (TRI) in the three-compartment electrochemical cell using a glassy carbon working electrode (2 mm of diameter) and a solution of ACN:H_2O 99:1 and 100 mM tetrabutylammonium perchlorate (TBAP). Potential scan rate was 50 mV s^{-1}. Arrows indicate the signal corresponding to the forward and reverse scans.

Electrolysis experiments were initially performed at a constant potential of 1050 mV, i.e., at a potential between the two oxidation peaks observed in the cyclic voltammetry experiments. The solution was stirred to maximize the supply of reagent to the surface of the electrode. During these experiments, a layer of precipitate formed on the surface of the working electrode regardless of the material used, Au or Pt. That solid product accumulated until small flakes started to detach and settle at the bottom of the cell. At the end of the electrolysis, the solution was diluted by a factor of 1000 in water and analyzed by LC-QqTOFMS. Results did not show the presence of any reaction product. To investigate the nature of the layer formed on the electrode, the cell was placed in a sonicator for 1 min to detach the precipitate from the glass wall and facilitate its recovery by dissolving it in methanol. Analysis of that methanol solution by LC-QqTOFMS showed only the presence of TRI. The presence of TRI in that solution was most likely the result of a contamination, i.e., TRI was adsorbed on the layer of precipitate.

When the potential applied during the electrolysis was increased to 1500 mV, i.e., after the second oxidation peak of TRI, and after 60 min of electrolysis, the solution acquired a slightly beige tint and no layer was formed on the surface of the electrode, thus suggesting the formation of soluble oxidation products, and a sample of the solution was collected. The sample was diluted by a factor of 1000 in water and injected in the LC-QqTOFMS. Results showed the presence of an oxidation product having an ion of m/z 307.1411 which was named oxidation product 306 or OP306 because its molecular mass is 306 Da. Analysis of spectral accuracy of the isotopic pattern in MassWorks indicated that only two neutral formulas were possible for that ion considering the constraints specified in the Material and methods section: $C_{14}H_{18}N_4O_4$ ($\Delta m = 1$ mDa, spectral accuracy = 93.9%) and $C_9H_{10}N_6O_6$ ($\Delta m = 5$ mDa, spectral accuracy = 92.2%). Since the number of N atoms in the oxidation should not be higher than in TRI and considering that the spectral accuracy for the first formula was higher, the unequivocal molecular formula was determined to be $C_{14}H_{18}N_4O_4$. This formula indicated the addition of one O atom relative to TRI ($C_{14}H_{18}N_4O_3$) in OP306. HDX experiments using a technique recently developed in our group [32] indicated an increase of one exchangeable H atom in OP306 (total of five exchangeable H atoms) relative to TRI (four exchangeable H atoms). MS/MS experiments with OP306 using a collision energy of 30 eV gave the product ions shown in Table 1. Transformation products of TRI with the same molecular formula, number of exchangeable hydrogens, precursor ion and product ions (Table 1) have been previously reported by several authors and were produced by rat metabolism [32], pig liver microsomes [33] and diverse water treatment processes [16,18,19,22]. Proposed structures are shown in Figure 2 (isomers A, B and C).

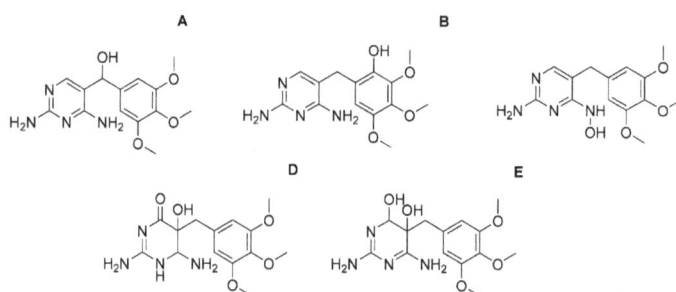

Figure 2. Tentative molecular structures for TRI transformation products according to previous studies. (**A–C**): Proposed structures for OP306. (**D,E**): Proposed structures for OP324.

From isomers A, B and C shown in Figure 2, OP306 appears to correspond to α-hydroxytrimethoprim (α-OH-TRI, isomer A) since it is the best match to MS/MS data. Two reaction mechanisms, based on single-electron transfer and nucleophilic attack by H_2O, are proposed to explain the electrochemical formation of OP306 (Figure 3 and Figure S1, Supplementary material). A hypothesis explaining the difference between the electrolysis products observed at 1050 and 1500 V is illustrated in Figure 3. The first step in the electrolysis of TRI, is the loss of an electron which generates a radical intermediate. This radical intermediate can react with TRI to form dimers or other TRI oligomers. At higher potentials, further oxidation (followed by the loss of a proton) of the radical intermediate is possible, thus resulting in a second cationic intermediate that cannot be further oxidized and that ultimately leads to the formation of α-OH-TRI after the loss of a proton and nucleophilic attack by water.

The formation of α-OH-TRI by electrochemistry in the electrolysis experiments was selective as is shown in Figure S2 (Supplementary material). No other major oxidation product between m/z 200 and 400 was observed in the chromatograms after 7 min of analysis (TRI eluted at around 6 min). While it is possible that dimers or other TRI dimers or oligomers could have been formed in the electrolysis at 1500 V, they do not appear to be major products. Also, detection of such species in the conditions used could be difficult since they would have eluted by the end of the chromatographic gradient. In those conditions, many column contaminants are eluted given the high organic content in the mobile phase. Those compounds can cause signal suppression, thus avoiding detection of minor sample components.

Table 1. Proposed molecular formulas and major product ions of the precursor ions observed in the electrochemistry experiments.

Observed Ions (m/z)	Name	Most Likely Molecular Formula [a] (Neutral)	Δm (mDa)	Spectral Accuracy (%)	RDBE [b]	Product Ions [c] (m/z)
Electrolysis						
307.1411	OP306	$C_{14}H_{18}N_4O_4$	1	93.9	7.5	259.0825 (100), 243.0875 (58), 274.1055 (22), 289.1286 (11)
Electro-assisted Fenton reaction						
237.1026	OP236	$C_{10}H_{12}N_4O_3$	4	64.6	6.5	N.A.
		$C_{13}H_{16}O_4$	−10	64.8	5.5	
291.1082	OP290	$C_{13}H_{14}N_4O_4$	0.5	N.A.	8.5	258.0736 (100), 273.0984 (66), 240.0645 (44), 241.0703 (40), 291.1084 (31)
307.1398	OP306	$C_{14}H_{18}N_4O_4$	−0.2	69.8	7.5	259.0822 (100), 243.0878 (60), 274.1054 (30), 231.0869 (14), 244.0927 (14)
323.1345	OP322a	$C_{14}H_{18}N_4O_5$	−0.5	89.2	7.5	249.0983 (100), 231.0887 (88), 259.0827 (86), 216.0624 (73), 323.1345 (59)
323.1346	OP322b	$C_{14}H_{18}N_4O_5$	−0.4	84.6	7.5	249.0979 (100), 231.0875 (90), 259.0826 (78), 216.0637 (45), 323.1349 (41)
325.1504	OP324	$C_{14}H_{20}N_4O_5$	−0.2	87.3	6.5	181.0680 (100), 325.1497 (55)
Thin later flow cell coupled online to high-resolution mass spectrometry (HRMS) [d]						
307.1421	OP306	$C_{14}H_{18}N_4O_4$	2	50.3	7.5	N.A.
323.1372	OP322	$C_{14}H_{18}N_4O_5$	2	92.9	7.5	323.1372 (100), 259.0998 (22), 291.1109 (20), 231.0893 (13)
339.1325	OP338	$C_{14}H_{18}N_4O_6$	3	N.A.	7.5	N.A.
357.1431	OP356	$C_{14}H_{20}N_4O_7$	3	N.A.	6.5	N.A.
398.1700	OP397	$C_{10}H_{23}N_9O_8$ *	−4	94.1	3.5	N.A.
		$C_{11}H_{23}N_7O_9$ *	7	94.1	3.5	
		$C_{16}H_{23}N_5O_7$ *	3	92.0	7.5	

[a] According to mass accuracy (tolerance = 10 mDa), spectral accuracy and number of C and N atoms in the candidate structures (candidates with number of C and N atoms higher than those of trimethoprim (TRI) were eliminated). [b] Ring and double bond equivalents. [c] Values between parentheses indicate normalized abundance at a collision energy of 30 eV. [d] Values between parentheses indicate normalized abundance at a collision energy of 20 eV. * Indicates that no possible candidate formula was possible within the constraints given, therefore a higher number of C and N atoms relative to TRI was allowed to determine a molecular formula. N.A.: Not available.

The TRI oxidation product α-OH-TRI has been observed in living systems such as rat metabolism [32], and biological and chemical process like nitrifying bacteria in activated sludge [22], direct photolysis and solar TiO_2 photocatalysis [16], oxidation by $KMnO_4$ [18] and thermo-activated persulfate oxidation [19]. This demonstrates the diversity of phenomena in which electrochemistry can play an important role in predicting, identifying and understanding the formation of transformation products of EOCs generated by diverse and dissimilar systems involving redox reactions.

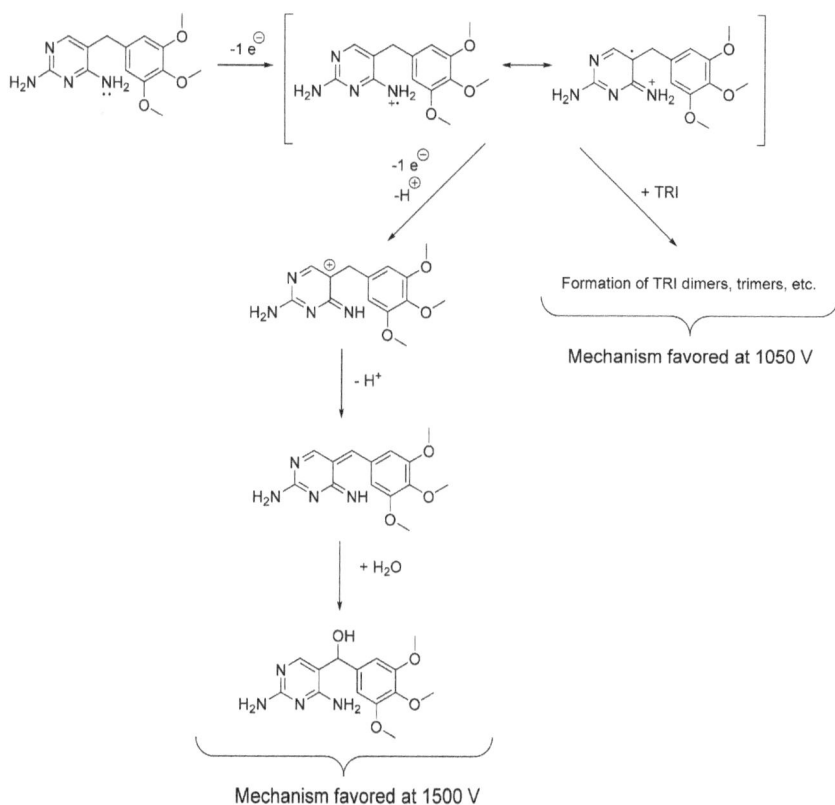

Figure 3. Proposed mechanism of formation of α-OH-TRI by electrolysis at 1500 V. The formation of the unidentified precipitated at 1050 V could be explained by the reaction of a radical intermediate with TRI.

3.2. Electrochemically-Assisted Fenton Reaction

These experiments showed that after 60 min of reaction, TRI was completely transformed (Figure 4). Several major transformation products with ions of m/z 237, 291, 307, 323 and 325 were formed during the experiment and they reached their maximum concentration between 10 to 20 min of treatment. After this maximum, the transformation products were also degraded and were no longer detected after 60 min of treatment, except for one transformation product which appeared to be resistant to degradation, the oxidation product with a m/z of 237.

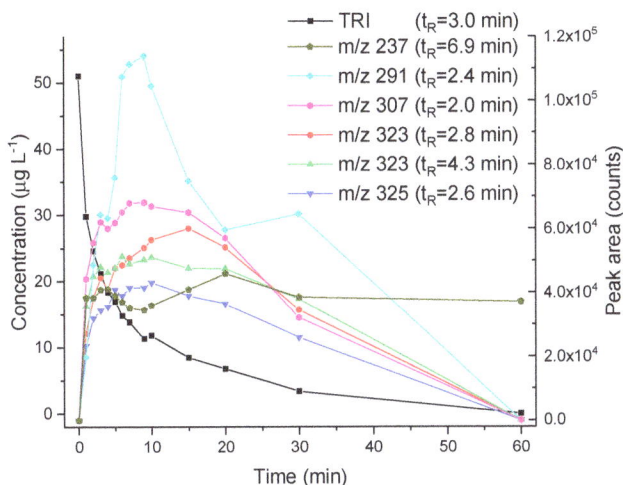

Figure 4. Profile of degradation of TRI and formation of its transformation products with the electro-assisted Fenton reaction setup. Only the concentration of TRI was measured (left axis). For the oxidation products, peak areas were used to measure their relative abundance as a function of time (right axis).

To the authors' knowledge, OP236 (m/z 237.1026) has not been reported previously in the literature. For this compound (Table 1), only two candidates were possible: $C_{10}H_{12}N_4O_3$ (Δm = 4 mDa, spectral accuracy = 64.6%), and $C_{13}H_{16}O_4$ (Δm = −10 mDa, spectral accuracy = 64.8%). Unfortunately, it was not possible to perform MS/MS experiments since precursor abundance was too low and further experiments are necessary to unambiguously assign its molecular formula.

As for m/z 291.1083 (OP290), formula determination based on mass and spectral accuracy suggested that one two molecular formulas are possible candidates: $C_{13}H_{14}N_4O_4$ (Δm = −0.5 mDa, spectral accuracy = 74.1%) or $C_{12}H_{18}O_8$ (Δm = 0.9 mDa, spectral accuracy = 74.1%). From those two, $C_{13}H_{14}N_4O_4$ is obviously the most probable candidate since it is extremely unlikely that TRI could have lost four N atoms while retaining 12 C atoms (loss of two C atoms) and accepted four additional O atoms upon its oxidation. Five potential structures corresponding to the formula $C_{13}H_{14}N_4O_4$ were suggested, as presented in Table 2. An in-silico MS/MS fragmentation analysis based on theoretical and library reactions on MassFrontier software showed that only one among those five structures (isomer E) could explain 10 of the 12 most abundant product ions of m/z 291.1083, with mass accuracy for all matching fragments ≤2.5 mDa.

In a previous study of the transformation of TRI under aerobic conditions in nitrifying activated sludge, this same structure was assigned to a transformation product of TRI with the same exact mass, albeit a completely different MS/MS spectrum [34]. While differences in MS/MS could be the result of different mass analyzers (the present study used an QqTOF while in the study of Jewell et al. a linear ion trap–orbitrap mass spectrometer was used) mass accuracy, spectral accuracy and comparison of MS/MS spectrum and in silico fragmentation suggests that the most likely structure of OP290 is isomer E or (2,4-diaminopyrimidin-5-yl)-(4-hydroxy-3,5-dimethoxyphenyl)methanone.

Table 2. Mass accuracy of the product ions of the five proposed structures for OP290 according to in-silico fragmentation analysis done by Mass Frontier software.

Observed Product Ions m/z	Relative Abundance %	Isomer				
		A	B	C	D	E
212.0700	10.5					−0.7
216.0519	14.3					
230.0790	18.3					0.8
234.0660	9.4					−2.5
240.0645	43.9					
241.0703	40.4		1.7			1.7
249.0866	25.4	11.6	0.4	0.4	11.6	0.4
258.0736	100	13.7	1.1	1.1	1.1	1.1
261.0613	8.0					0.5
261.0968	6.5	1.4	1.4	1.4	1.4	1.4
273.0984	66.4	−0.2	−0.2	−0.2	−0.2	−0.2
276.0849	15.0	13.0	13.0	13.0	13.0	0.4

Note: A few ions were omitted from this table if the accurate mass and the abundance were close to the expected M + 1 isotopic peaks.

Chromatograms showed that the oxidation product with a retention time of 2.0 min had an ion of m/z 307.1398 and eluted earlier than TRI on the C_{18} column, similar to OP306 (α-OH-TRI), the oxidation product observed in the electrolysis experiments with the three-compartment cell with stationary electrodes. Mass accuracy (-0.2 mDa) and spectral accuracy (69.8%) also suggested that the most likely neutral formula for that compound is $C_{14}H_{18}N_4O_4$ since the other two possible candidates ($C_{16}H_{14}N_6O$, $C_9H_{18}N_6O_6$) had a higher number of N atoms relative to TRI, which is not possible in this reactive system. The low spectral accuracy of the most likely candidate was due to the low signal-to-noise ratio signal, and the presence of background peaks interfering with the determination of spectral accuracy. Peaks in the MS/MS spectrum of m/z 307.1398 (Table 1) showed that product ions generated after collision induced dissociation (m/z 243, m/z 244, m/z 259 are m/z 274) are the same as those of OP306 observed in the three-compartment electrode cell with stationary electrodes (Table 1). This suggests that α-OH-TRI can also be formed in the electro-assisted Fenton reaction setup.

The presence of two oxidation products of m/z 323, OP322a eluting at 2.8 min, and the other, OP322b, with a retention time of 4.3 min was also observed. Spectral accuracies (89.2% and 84.6%, respectively) and mass accuracies (-0.5 and -0.4 mDa, respectively) unambiguously indicated that the neutral formula of both OP322a and OP322b was $C_{14}H_{18}N_4O_5$. MS/MS experiments revealed that those two isomers share the same major product ions: m/z 249, m/z 231, m/z 259, m/z 216 and m/z 232 (Table 1). Therefore, they must have a very similar structure. For example, the addition of oxygen must have occurred at close positions on the OP322a and OP322b molecule, such as on the ring of the 3,4,5-trimethoxyphenyl moiety or on the amine groups to yield N-oxides. However, those oxygen additions decreased the retention of OP322a more significantly than that of OP322b, since the difference of retention time between the two isomers is of 1.5 min. Oxidation products of TRI with the same molecular formula have been reported earlier by several authors working on advanced oxidation processes for water treatment and these were identified as dihydroxy-TRI isomers [16,19]. At least five isomers of OP322 were reported by Sirtori, et al. [16]. In another study, two other isomers were proposed for OP322 and were explained by cleavage of the 2,4-diaminopyrimidinyl moiety but the MS/MS spectrum reported by the authors in that study [18] had only one product ion (m/z 181) which does not correspond the one observed experimentally in the present study (Table 1).

Another oxidation product observed was m/z 325.1504 (OP324). Spectral accuracy (87.3%) and mass accuracy (-0.2 mDa) pointed to $C_{14}H_{20}N_4O_5$ as the most likely molecular formula and the other possible candidates (e.g., $C_{20}H_{20}O_4$ or $C_{16}H_{16}N_6O_2$) had a higher number of C or N atoms relative to TRI. A compound with the same molecular formula (compound D, Figure 2) was also identified in TRI degradation experiments with sludge containing nitrifying bacteria [22]. The only product ion of m/z 325.1504 observed by MS/MS, m/z 181 (Table 1), is also the only product ion reported by Eichhorn et al. [22]. Another study on the oxidation of antibiotics during water treatment with $KMnO_4$ [18] suggested a different molecular structure for a TRI oxidation product of formula $C_{14}H_{20}N_4O_5$ and similar MS/MS spectrum. Such structure (Figure 2, compound E) is also possible. Unfortunately, HDX combined with MS/MS experiments were not performed on OP324, which could have helped determine the correct structure since some exchangeable hydrogens in both compounds are located in different parts of their structure.

Contrary to the CV and electrolysis experiments, in the electro-assisted Fenton reaction there was no interaction between TRI and the surface of the electrode. The reactions leading to these transformation products take place within the solution. Besides being able to mimic certain reactions resulting from the metabolism of cytochrome P450 enzymes, as reported earlier [10], the electro-assisted Fenton reaction is an interesting approach to study oxidation reactions occurring during water treatment processes involving OH•.

3.3. Thin-Layer Flow Cell Coupled Online to High-Resolution Mass Spectrometry

One of the major advantages of the thin-layer flow cell compared to the other setups used is the possibility of online coupling with a mass spectrometer. Such a setup allows us to detect in real time compounds formed in the cell when a potential is applied. This setup saves enormous amounts of time but it also comes with certain limitations: (i) the solution composition must be compatible with electrospray ionization, i.e., the supporting electrolyte must be kept at low concentration, usually around 0.1% v/v or ≤ 50 mM; (ii) the supporting electrolyte must be volatile, such as FA or ammonium formate; and (iii) there is no chromatographic data, reaction products are introduced directly into the mass spectrometer, therefore isomers cannot be resolved.

Figure 5 shows the results of experiments performed using constant potential steps with a solution containing TRI at 1.25 mg·L^{-1}. Preliminary experiments showed that no reaction product was observed below 750 mV (vs. H$_2$/Pd) using the boron-doped diamond (BDD) working electrode. Reaction products started to be detectable at +1000 mV and a progressive diminution of the TRI ion (m/z 291) was observed when increasing the potential. At +2500 mV, the signal of TRI was about 0.5% of its original value when the cell was off.

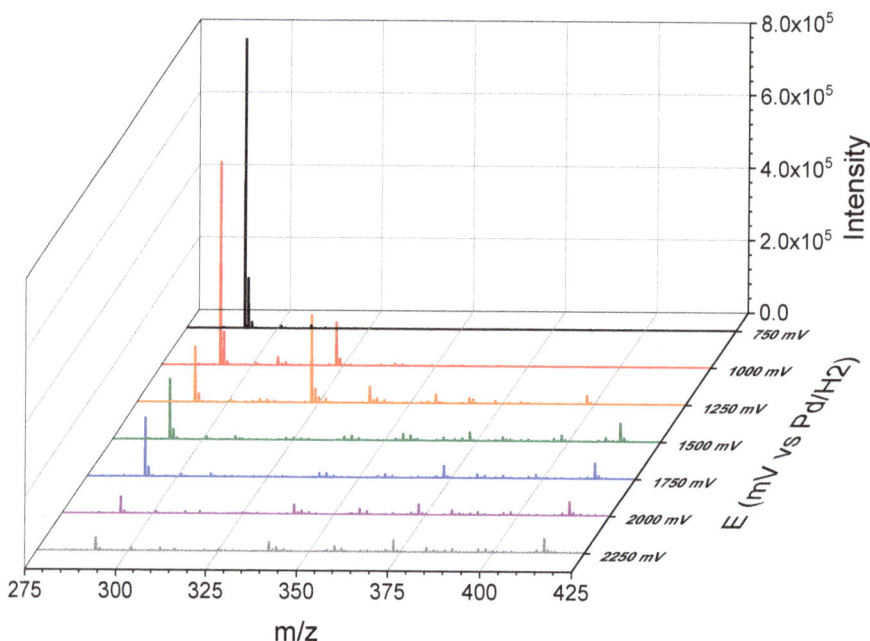

Figure 5. Mass spectrometry voltammogram of TRI obtained with the thin-layer flow cell coupled online with the QqTOFMS. The solution was composed of 0.1% FA in ACN:H$_2$O 1:1 and the concentration of TRI (m/z 291) was 1.25 mg·L^{-1}.

The most abundant oxidation products were detected between +1000 and +1250 mV and are reported in Table 1. Amongst that list, the most intense signal corresponded to m/z 323.1372 (C$_{14}$H$_{18}$N$_4$O$_5$, Δm = 2 mDa, spectral accuracy = 92.9%) and it was observed at both 1000 and 1250 mV (vs. Pd/H$_2$). Tandem mass (MS/MS) spectra experiments (Table 1) confirmed that C$_{14}$H$_{18}$N$_4$O$_5$ (or most likely a mixture of compounds of molecular formula C$_{14}$H$_{18}$N$_4$O$_5$) generated in the thin-layer flow cell are the same as the dihydroxy-TRI isomers (OP322a and OP322b) observed in the Electro-assisted Fenton reaction experiments. Also, m/z 307.1421 (C$_{14}$H$_{18}$N$_4$O$_4$, Δm = 2 mDa, spectral accuracy = 50.3%) observed only at 1000 mV (vs. H$_2$-Pd) and m/z 339.1325 (C$_{14}$H$_{18}$N$_4$O$_6$, Δm = 2 mDa) observed at 1250 mV (vs. H$_2$/Pd) were observed and appear to be mono and trihydroxylated-TRI, respectively. As for m/z 357.1431 (1250 mV vs. H$_2$/Pd), it could correspond to C$_{14}$H$_{20}$N$_4$O$_7$ (Δm = 3 mDa). However, unequivocal molecular formulas for this ion cannot be confirmed since the relative low abundance and signal-to-noise ratio of these compounds resulted in low spectral accuracy or isotope patterns that did not correspond to the proposed formulas, as was the case of m/z 339.1325 (1250 mV vs. H_2/Pd). Also, meaningful MS/MS spectra could not be obtained because of the low signal of these two ions. However, the presence of multiple hydroxylated oxidation products of TRI in the thin-layer flow cell is highly possible. Multiple hydroxylated species of nucleotides have been observed when using a similar setup with the same electrode, albeit at both higher pH and potential [35]. Electron transfer from TRI to the electrode surface may not be the only mechanism responsible for the formation of TRI oxidation products in the BBD electrode; OH• radicals most probably intervened in their formation as well. Indeed, Marselli, et al. [36] have shown that OH• radicals are formed in the oxidation of water on a BDD electrode. A previous study on the degradation of TRI in a photoelectro-Fenton process also reported the formation of multiple hydroxylated oxidation products of TRI [37].

The presence of m/z 398.1700 (1250 mV vs. H_2/Pd), is however puzzling since no molecular formula with a number of C and N atoms equal or lower than those in TRI can be assigned to this ion. Therefore, this ion may correspond to an addition or condensation reaction product between TRI oxidation products. Such reactions are not unlikely, since the generation of radical species by electrochemistry often leads to oligomerization or polymerisation reactions [38]. Studies on the oxidation of the skin allergen *p*-phenylenediamine using a thin-layer flow cell with a BBD working electrode showed that ions corresponding to *p*-phenylenediamine dimers were formed [39]. The nature of such reactions products was not investigated here and were considered out of the scope of the present study.

Results obtained with this setup showed that coupling a thin-layer flow cell with a BDD electrode to a high-resolution mass spectrometer provides an interesting tool for studying the oxidation products of organic compounds in the environment and water treatment process involving OH• such as photolysis and advanced oxidation processes. In this setup, the effect of potential on the formation of oxidation products can be monitored in real time, which allows a more efficient interpretation of data.

In summary, the diverse electrochemical approaches used in the present study showed that mostly mono-, di-hydroxylated and possibly tri-hydroxylated derivatives of trimethoprim were generated electrochemically (Table 3). Those compounds have been previously reported as mammalian [32] and bacterial metabolites [22] as well as transformation products of advance oxidation processes applied to waters containing trimethoprim [16,18,19,37]. These results suggest that electrochemistry–high resolution mass spectrometry is an interesting technique for studying oxidative reactions of organic compounds of environmental interest, as was previously suggested by Hoffmann et al. [12].

Table 3. Summary of techniques tested and the main results obtained.

Technique	Conditions	TRI Oxidation Products Generated	Identification Level [29]	Previous Reports of the Oxidation Product
Electrolysis	*Solution:* ACN:H$_2$O 99: 1 (*v/v*) with 100 mM TBAP. *WE:* Glassy carbon maintained under a N$_{2(g)}$ atmosphere. *Potential applied:* 1500 mV (vs. Ag/Ag$^+$).	OP306 (α-OH-TRI)	Probable structure (level 2b): based on spectral and mass accuracy, H/D exchange, experimental and literature MS/MS spectra.	Rat metabolism [32]. Nitrifying bacteria in activated sludge [22], direct photolysis and solar TiO$_2$ photocatalysis [16], Oxidation by KMnO4 [18], thermo-activated persulfate oxidation [19].
Electro-assisted Fenton reaction	*Solution:* 50 mM Na$_2$SO$_4$, 0.1 mM FeSO$_4$ in acidified H$_2$O with H$_2$SO$_4$ at pH 2. *WE:* Glassy carbon. *Current density:* 1 mA cm^{-2}.	OP236 (*m/z* 237.1026)	Accurate mass (level 5): mass and spectral accuracy could not assign unequivocally a formula to the observed *m/z*.	Not reported previously.
		OP290 [2,4-diaminopyrimidin-5-yl)-(4-hydroxy-3,5-dimethoxyphenyl) methanone]	Probable structure (level 2b): based on spectral and mass accuracy, and comparative in-silico MS/MS fragmentation analysis.	Nitrifying bacteria in activated sludge [34].
		OP306 (α-OH-TRI)	Probable structure (level 2b): based on spectral and mass accuracy, experimental and literature MS/MS spectra.	Same as indicated for the three-compartment cell with stationary electrodes.
		OP322 (2OH-TRI) isomers	Tentative structures (level 3): based on spectral and mass accuracy, experimental and literature MS/MS spectra.	Direct photolysis and solar TiO$_2$ photocatalysis [16], thermo-activated persulfate oxidation [19].
		OP324 (C$_{14}$H$_{20}$N$_4$O$_5$)	Tentative structures (level 3): based on spectral and mass accuracy. Experimental and literature MS/MS spectra could not assign unambiguously one structure.	Nitrifying bacteria in activated sludge [22].
Thin-layer flow cell coupled online to HRMS	*Solution:* 0.1% FA in H$_2$O:ACN 1:1 *WE:* Boron-doped diamond. *Potential applied:* 1000 to 1500 vs. Pd/H$_2$.	OP306 (*m/z* 323.1372)	Accurate mass (level 5): mass and spectral accuracy could not assign unequivocally a formula to the observed *m/z*.	Same as indicated for the three-compartment cell with stationary electrodes.
		OP322 (2OH-TRI)	Tentative structures (level 3): based on spectral and mass accuracy, experimental and literature MS/MS spectra.	Same as indicated for the three-compartment cell with stationary electrodes.
		OP338 (*m/z* 339.1325)	Exact mass (level 5): mass and spectral accuracy could not assign unequivocally a formula to the observed *m/z*.	Photoelectro-Fenton with Pt anode [37].

WE: Working electrode.

4. Conclusions

Oxidation experiments with TRI using different electrochemical experiments (cyclic voltammetry, electrolysis, electro-assisted Fenton reaction and thin-layer flow cell coupled online to HRMS) generated several oxidation products previously reported in diverse biological processes such as bacterial and mammalian metabolism as well as in oxidation processes used for water treatment. Since many parameters intervened in the outcome of the experiments with each technique such as solution composition, pH, and electrode material, it is not possible to rank the techniques tested in terms of performance. Table 2 shows that the four setups differed especially in terms of selectivity in the production of oxidation products. Electrolysis using a glassy carbon electrode under $N_{2(g)}$ atmosphere was the most selective. This setup only generated OP306, identified as α-hydroxytrimethoprim (α-OH-TRI), as the major oxidation product. Electro-assisted Fenton oxidation and oxidation at a BDD anode in a thin-layer flow cell were less selective and generated an array of oxidation products. Among the most abundant were α-OH-TRI, 2OH-TRI and OP338, possibly a tri-hydroxylated derivative of TRI. The selectivity depends on the main oxidation mechanism involved in the techniques evaluated. In the three-compartment cell, the formation of α-OH-TRI is initiated by the transfer of one electron to the working electrode. In the electro-assisted Fenton oxidation in the one-compartment cell and in the oxidation at a BDD anode in the thin-layer flow cell, the OH• radical, a highly reactive oxidant with low selectivity towards organic compounds, plays a major role. Also, further studies with the thin-layer flow cell using different working electrodes, solution compositions, and pH values are needed, since the rapidity and simplicity of this technique makes it a quick and simple approach to study the fate of EOCs.

Finally, this study confirmed that electrochemical techniques are relevant not only to mimic the cytochrome P450 oxidation transformations of drugs, as has been suggested previously [10,13], but also to study the oxidation reactions of organic contaminants in wastewater treatment plants. The key role that redox reactions play in the environment make electrochemistry coupled to high resolution mass spectrometry a powerful technique to improve our understanding of the fate of EOCs in the environment.

Supplementary Materials: The following are available online at www.mdpi.com/2076-3298/5/1/18/s1, Figure S1. Simultaneous proposed mechanism of formation of α-OH-TRI (OP306). Since protons are generated at the anode during the electrolysis, TRI can be protonated in solution and then oxidized by initial loss of an electron from the trimethoxyl moiety rather than the diaminopyridinyl moiety. Figure S2. Survey view of the chromatogram obtained by UPLC-QTOFMS of a solution of TRI after 60 min of electrolysis at 1500 mV vs. Ag/Ag+ using the setup described for the three-compartment electrochemical cell with stationary electrodes. The red circle indicates the peak corresponding to OP306 (m/z 307, α-hydroxytrimethoprim) and the blue square the peak corresponding to TRI (m/z 291). Peaks observed after 7 min correspond to compounds eluted at high organic percentage during the chromatographic separation. Signal threshold for the survey view was 1000 counts.

Acknowledgments: This research was funded by NSERC through a Discovery grant to Pedro A. Segura.

Author Contributions: Marc-André Lecours, Jean Lessard, Gessie M. Brisard and Pedro A. Segura conceived and designed the experiments; Marc-André Lecours performed the experiments and data analysis; Emmanuel Eysseric performed the spectral accuracy analysis; Viviane Yargeau contributed with access to Mass Frontier software; Jean Lessard provided mechanisms of oxidation of trimethoprim; Marc-André Lecours and Pedro A. Segura wrote the paper. All authors participated in the revision of the manuscript.

Conflicts of Interest: There are no conflicts of interest to declare.

References

1. Snyder, M.J. Cytochrome P450 enzymes in aquatic invertebrates: Recent advances and future directions. *Aquat. Toxicol.* **2000**, *48*, 529–547. [CrossRef]
2. Lohmann, W.; Karst, U. Biomimetic modeling of oxidative drug metabolism. *Anal. Bioanal. Chem.* **2008**, *391*, 79–96. [CrossRef] [PubMed]
3. Jahn, S.; Karst, U. Electrochemistry coupled to (liquid chromatography/) mass spectrometry—Current state and future perspectives. *J. Chromatogr. A* **2012**, *1259*, 16–49. [CrossRef] [PubMed]

4. Yan, Z.; Caldwell, G.W. Stable-isotope trapping and high-throughput screenings of reactive metabolites using the isotope MS signature. *Anal. Chem.* **2004**, *76*, 6835–6847. [CrossRef] [PubMed]
5. Glaze, W.H.; Kang, J.-W.; Chapin, D.H. The chemistry of water treatment processes involving ozone, hydrogen peroxide and ultraviolet radiation. *Ozone-Sci. Eng.* **1987**, *9*, 335–352. [CrossRef]
6. Ikehata, K.; Naghashkar, N.J.; El-Din, M.G. Degradation of aqueous pharmaceuticals by ozonation and advanced oxidation processes: A review. *Ozone-Sci. Eng.* **2006**, *28*, 353–414. [CrossRef]
7. Segura, P.A.; Saadi, K.; Clair, A.; Lecours, M.-A.; Yargeau, V. Application of XCMS online and toxicity bioassays to the study of transformation products of levofloxacin. *Water Sci. Technol.* **2015**, *72*, 1578–1587. [CrossRef] [PubMed]
8. Boxall, A.; Rudd, M.; Brooks, B.; Caldwell, D.; Choi, K.; Hickmann, S.; Innes, E.; Ostapyk, K.; Staveley, J.; Verslycke, T.; et al. Pharmaceuticals and Personal Care Products in the Environment: What are the Big Questions? *Environ. Health Perspect.* **2012**, *120*, 1221–1229. [CrossRef] [PubMed]
9. Bussy, U.; Chung-Davidson, Y.-W.; Li, K.; Li, W. Phase I and phase II reductive metabolism simulation of nitro aromatic xenobiotics with electrochemistry coupled with high resolution mass spectrometry. *Anal. Bioanal. Chem.* **2014**, *406*, 7253–7260. [CrossRef] [PubMed]
10. Johansson, T.; Weidolf, L.; Jurva, U. Mimicry of phase I drug metabolism–novel methods for metabolite characterization and synthesis. *Rapid Commun. Mass Spectrom.* **2007**, *21*, 2323–2331. [CrossRef] [PubMed]
11. Jurva, U.; Wikström, H.V.; Weidolf, L.; Bruins, A.P. Comparison between electrochemistry/mass spectrometry and cytochrome P450 catalyzed oxidation reactions. *Rapid Commun. Mass Spectrom.* **2003**, *17*, 800–810. [CrossRef] [PubMed]
12. Hoffmann, T.; Hofmann, D.; Klumpp, E.; Küppers, S. Electrochemistry-mass spectrometry for mechanistic studies and simulation of oxidation processes in the environment. *Anal. Bioanal. Chem.* **2011**, *399*, 1859–1868. [CrossRef] [PubMed]
13. Bussy, U.; Boujtita, M. Advances in the electrochemical simulation of oxidation reactions mediated by cytochrome P450. *Chem. Res. Toxicol.* **2014**, *27*, 1652–1668. [CrossRef] [PubMed]
14. Seiwert, B.; Golan-Rozen, N.; Weidauer, C.; Riemenschneider, C.; Chefetz, B.; Hadar, Y.; Reemtsma, T. Electrochemistry Combined with LC–HRMS: Elucidating Transformation Products of the Recalcitrant Pharmaceutical Compound Carbamazepine Generated by the White-Rot Fungus *Pleurotus ostreatus*. *Environ. Sci. Technol.* **2015**, *49*, 12342–12350. [CrossRef] [PubMed]
15. Segura, P.A.; Takada, H.; Correa, J.A.; El Saadi, K.; Koike, T.; Onwona-Agyeman, S.; Ofosu-Anim, J.; Sabi, E.B.; Wasonga, O.V.; Mghalu, J.M.; et al. Global occurrence of anti-infectives in contaminated surface waters: Impact of income inequality between countries. *Environ. Int.* **2015**, *80*, 89–97. [CrossRef] [PubMed]
16. Sirtori, C.; Agüera, A.; Gernjak, W.; Malato, S. Effect of water-matrix composition on Trimethoprim solar photodegradation kinetics and pathways. *Water Res.* **2010**, *44*, 2735–2744. [CrossRef] [PubMed]
17. Kuang, J.; Huang, J.; Wang, B.; Cao, Q.; Deng, S.; Yu, G. Ozonation of trimethoprim in aqueous solution: Identification of reaction products and their toxicity. *Water Res.* **2013**, *47*, 2863–2872. [CrossRef] [PubMed]
18. Hu, L.; Stemig, A.M.; Wammer, K.H.; Strathmann, T.J. Oxidation of antibiotics during water treatment with potassium permanganate: Reaction pathways and deactivation. *Environ. Sci. Technol.* **2011**, *45*, 3635–3642. [CrossRef] [PubMed]
19. Ji, Y.; Xie, W.; Fan, Y.; Shi, Y.; Kong, D.; Lu, J. Degradation of trimethoprim by thermo-activated persulfate oxidation: Reaction kinetics and transformation mechanisms. *Chem. Eng. J.* **2016**, *286*, 16–24. [CrossRef]
20. Michael, I.; Hapeshi, E.; Osorio, V.; Perez, S.; Petrovic, M.; Zapata, A.; Malato, S.; Barceló, D.; Fatta-Kassinos, D. Solar photocatalytic treatment of trimethoprim in four environmental matrices at a pilot scale: Transformation products and ecotoxicity evaluation. *Sci. Total Environ.* **2012**, *430*, 167–173. [CrossRef] [PubMed]
21. Radjenović, J.; Godehardt, M.; Hein, A.; Farré, M.; Jekel, M.; Barceló, D. Evidencing generation of persistent ozonation products of antibiotics roxithromycin and trimethoprim. *Environ. Sci. Technol.* **2009**, *43*, 6808–6815. [CrossRef] [PubMed]
22. Eichhorn, P.; Ferguson, P.L.; Pérez, S.; Aga, D.S. Application of ion trap-MS with H/D exchange and QqTOF-MS in the identification of microbial degradates of trimethoprim in nitrifying activated sludge. *Anal. Chem.* **2005**, *77*, 4176–4184. [CrossRef] [PubMed]

23. Zhang, Z.; He, L.; Lu, L.; Liu, Y.; Dong, G.; Miao, J.; Luo, P. Characterization and quantification of the chemical compositions of Scutellariae Barbatae herba and differentiation from its substitute by combining UHPLC–PDA–QTOF–MS/MS with UHPLC–MS/MS. *J. Pharm. Biomed. Anal.* **2015**, *109*, 62–66. [CrossRef] [PubMed]

24. Nouri-Nigjeh, E.; Permentier, H.P.; Bischoff, R.; Bruins, A.P. Lidocaine oxidation by electrogenerated reactive oxygen species in the light of oxidative drug metabolism. *Anal. Chem.* **2010**, *82*, 7625–7633. [CrossRef] [PubMed]

25. Nouri-Nigjeh, E.; Permentier, H.P.; Bischoff, R.; Bruins, A.P. Electrochemical oxidation by square-wave potential pulses in the imitation of oxidative drug metabolism. *Anal. Chem.* **2011**, *83*, 5519–5525. [CrossRef] [PubMed]

26. Oloman, C.; Watkinson, A. Hydrogen peroxide production in trickle-bed electrochemical reactors. *J. Appl. Electrochem.* **1979**, *9*, 117–123. [CrossRef]

27. Dodd, M.C.; Buffle, M.-O.; Von Gunten, U. Oxidation of antibacterial molecules by aqueous ozone: Moiety-specific reaction kinetics and application to ozone-based wastewater treatment. *Environ. Sci. Technol.* **2006**, *40*, 1969–1977. [CrossRef] [PubMed]

28. Antec. *m-Prepcell User Manual (204.0010, 5th ed.)*; Antec: Zoeterwoude, The Netherlands, 2013; p. 45.

29. Schymanski, E.L.; Jeon, J.; Gulde, R.; Fenner, K.; Ruff, M.; Singer, H.P.; Hollender, J. Identifying small molecules via high resolution mass spectrometry: Communicating confidence. *Environ. Sci. Technol.* **2014**, *48*, 2097–2098. [CrossRef] [PubMed]

30. Eysseric, E.; Bellerose, X.; Lavoie, J.-M.; Segura, P.A. Post-column hydrogen-deuterium exchange technique to assist in the identification of small organic molecules by mass spectrometry. *Can. J. Chem.* **2016**, *94*, 781–787. [CrossRef]

31. Wang, Y.; Gu, M. The concept of spectral accuracy for MS. *Anal. Chem.* **2010**, *82*, 7055–7062. [CrossRef] [PubMed]

32. Meshi, T.; Sato, Y. Studies on sulfamethoxazole/trimethoprim. Absorption, distribution, excretion and metabolism of trimethoprim in rat. *Chem. Pharm. Bull.* **1972**, *20*, 2079–2090. [CrossRef] [PubMed]

33. Liu, W.-T.; Li, K.-C. Application of reutilization technology to calcium fluoride sludge from semiconductor manufacturers. *J. Air Waste Manag. Assoc.* **2011**, *61*, 85–91. [CrossRef] [PubMed]

34. Jewell, K.S.; Castronovo, S.; Wick, A.; Falås, P.; Joss, A.; Ternes, T.A. New insights into the transformation of trimethoprim during biological wastewater treatment. *Water Res.* **2016**, *88*, 550–557. [CrossRef] [PubMed]

35. Baumann, A.; Lohmann, W.; Jahn, S.; Karst, U. On-Line Electrochemistry/Electrospray Ionization Mass Spectrometry (EC/ESI-MS) for the Generation and Identification of Nucleotide Oxidation Products. *Electroanalysis* **2010**, *22*, 286–292. [CrossRef]

36. Marselli, B.; Garcia-Gomez, J.; Michaud, P.-A.; Rodrigo, M.; Comninellis, C. Electrogeneration of hydroxyl radicals on boron-doped diamond electrodes. *J. Electrochem. Soc.* **2003**, *150*, D79–D83. [CrossRef]

37. Moreira, F.C.; Garcia-Segura, S.; Boaventura, R.A.; Brillas, E.; Vilar, V.J. Degradation of the antibiotic trimethoprim by electrochemical advanced oxidation processes using a carbon-PTFE air-diffusion cathode and a boron-doped diamond or platinum anode. *Appl. Catal. B Environ.* **2014**, *160*, 492–505. [CrossRef]

38. Gattrell, M.; Kirk, D. A study of electrode passivation during aqueous phenol electrolysis. *J. Electrochem. Soc.* **1993**, *140*, 903–911. [CrossRef]

39. Jahn, S.; Faber, H.; Zazzeroni, R.; Karst, U. Electrochemistry/mass spectrometry as a tool in the investigation of the potent skin sensitizer p-phenylenediamine and its reactivity toward nucleophiles. *Rapid Commun. Mass Spectrom.* **2012**, *26*, 1453–1464. [CrossRef] [PubMed]

environments

MDPI

Article
Effects of Physico-Chemical Post-Treatments on the Semi-Continuous Anaerobic Digestion of Sewage Sludge

Xinbo Tian [1] and Antoine Trzcinski [2,*]

[1] Advanced Environmental Biotechnology Centre, Nanyang Environment and Water Research Institute, Nanyang Technological University, 1 Cleantech Loop, Singapore 637141, Singapore; TIAN0047@e.ntu.edu.sg
[2] Faculty of Health, Engineering and Sciences, School of Civil Engineering & Surveying, University of Southern Queensland, Toowoomba 4350, Australia
*   Correspondence: antoine.trzcinski@usq.edu.au; Tel.: +61-(0)7-4631-1617

Received: 20 June 2017; Accepted: 12 July 2017; Published: 13 July 2017

Abstract: Sludge production in wastewater treatment plants is increasing worldwide due to the increasing population. This work investigated the effects of ultrasonic (ULS), ultrasonic-ozone (ULS-Ozone) and ultrasonic + alkaline (ULS+ALK) post-treatments on the anaerobic digestion of sewage sludge in semi-continuous anaerobic reactors. Three conditions were tested with different hydraulic retention times (HRT, 10 or 20 days) and sludge recycle ratios (R = Q_R/Q_{in} (%): 50 or 100%). Biogas yield increased by 17.8% when ULS+ALK post-treatment was applied to the effluent of a reactor operating at 20 days HRT and at a 100% recycle ratio. Operation at 10 days HRT also improved the biogas yield (277 mL CH_4/g VS_{added} (VS: volatile solids) versus 249 mL CH_4/g VS_{added} in the control). The tested post-treatment methods showed 4–7% decrease in effluent VS. The post-treatment resulted in a decrease in the cellular ATP (Adenosine tri-phosphate) concentration indicating stress imposed on microorganisms in the reactor. Nevertheless, this did not prevent higher biogas production. Based on the results, the post-treatment of digested sludge or treating the sludge between two digesters is an interesting alternative to pre-treatments.

Keywords: sewage sludge; ultrasound; ozone; post-treatment; anaerobic digestion; biogas

1. Introduction

Hydrolysis of particulate organics is known to limit the rate of sludge anaerobic digestion [1,2]. Pre-treatment has been widely reported to solubilize the organic solids in sludge and make them more accessible for the subsequent anaerobic digestion [3–5]. Ultrasonication (ULS) (20 kHz) pre-treatment at 6250 and 9350 kJ/kg total solids (TS) resulted, respectively, in a 47% and 51% increase in methane production [6]. Anaerobic biodegradability of feed sludge after combined ultrasound (ULS) and ozone pre-treatment increased by 93% and 106% after 30 min and 45 min [7]. Combined ULS and alkaline (ALK) treatment of feed sludge enhanced the subsequent anaerobic digestion performance. Kim et al. [8] found that the methane production improved by around 55% along with 17% increase in volatile solids (VS) reduction after combined ULS and ALK pre-treatment (pH9 + 7000 kJ/kg TS). These studies were carried out in batch mode which often does not reflect the performance of full-scale continuous digesters.

Post-treatment is realized by treating the digested sludge and recycling the treated digested sludge back to the original anaerobic reactor as shown in Figure 1. The concept was first proposed by Gossett et al. [9] who found the performance of thermal treatment was more efficient when the substrate (i.e., municipal refuse) was first biodegraded compared to the situation where the thermal treatment was directly applied to the substrate. Pre-treatment of sludge before AD is often applied to solubilize these solids to accelerate subsequent digestion. The rationale for post-treatment and pre-treatment is similar wherein both aim to rupture the microbial cells and release the extra- and intra-cellular substances. However, Takashima et al. [10] indicated pre-treatment not only targets the slowly biodegradable solids, but also the easily biodegradable solids in waste activated sludge (WAS). As a result, part of the energy and chemical input during pre-treatment would then be wasted on solubilizing the easily biodegradable organic particulates without increasing overall sludge biodegradability. Takashima et al. [10] suggested that the post-treatment of digested sludge and recycling the treated digested sludge back to the anaerobic reactor could be an alternative to pre-treatment.

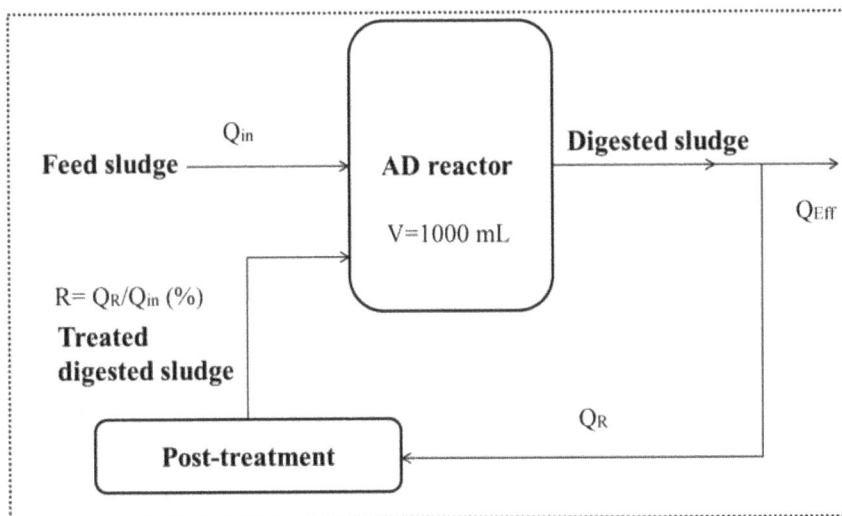

Figure 1. Schematic diagram of the anaerobic digestion process incorporating a post-treatment. Q_{in} = influent flowrate (mL/day), Q_{eff} = effluent flowrate (mL/day), Q_R = recycled flowrate (mL/day), and V = working volume (mL).

As the digested sludge contains primarily slowly biodegradable and refractory solids, the energy of the post-treatment focuses on converting the non-biodegradable solids into biodegradable ones [10–12].

In some wastewater treatment plants, the highly-biodegradable primary sludge (PS) and more recalcitrant waste activated sludge (WAS) are combined. In this situation, the post-treatment could be more efficient than the pre-treatment, because the solids in PS contain a higher content of biodegradable solids [13]. Compared to pre-treatment, post-treatment would more specifically target the solids which are more difficult to be biodegraded in digested sludge.

However, the studies on post-treatment techniques are relatively scarce in comparison with the information on pre-treatment and most studies carried out batch anaerobic digestion tests. Ozone [14,15], alkaline [12], thermal [13,16], thermal/acid [17,18], and thermal/alkaline [11] was successfully applied for sludge post-treatment. There are also papers showing post-treatment were superior to pre-treatment in terms of improving the anaerobic digestion effectiveness [13,16]. Battimelli et al. [14] and Li et al. [12] indicated the recycle ratio of the post-treated sludge to be an important operation parameter. It has impacts on the actual solids retention times (SRT) of the anaerobic reactor, as well as the anaerobic digestion performance.

However, ultrasonic (ULS), combined ultrasonic-ozone (ULS-Ozone), and combined ultrasonic with alkaline (ULS+ALK) post-treatments have not yet been documented in continuous reactors. Accordingly, information about the anaerobic digestion performance and the stress on microbial communities with post-treatment at different hydraulic retention times (HRTs) is not available.

Therefore, this work aims to compare the influence of the ULS, ULS-Ozone and ULS+ALK post-treatments on the anaerobic digestion performance of sewage sludge in semi-continuous reactors, as well as comparing the performance of pre- and post-treatment under the same conditions. The change in daily biogas production and suspended solids concentration were used to evaluate the anaerobic digestion performance at different HRTs and recycle ratios.

2. Materials and Methods

2.1. Sludge Sample

Sludge consisting of a mixture of primary sludge and thickened waste activated sludge (ratio 1:1 based on dry solids, TS: 15.2 ± 0.4 g/L) were collected from a local municipal wastewater reclamation plant. Fresh sludge was collected on four occasions and the main parameters were measured in triplicate for each sludge batch. Due to this, the solids content varied slightly between batches and a range of values is provided in Table 1.

Table 1. Characteristics of mixed sludge. The range of values in four consecutive batches of sludges for which each analyses was conducted in triplicate. COD: chemical oxygen demand.

Parameter	Value Range
Total Solids (g/L)	14.8–15.6
Volatile Solids (g/L)	12.1–13.3
Total Suspended Solids (g/L)	13.1–13.9
Volatile Suspended Solids (g/L)	10.5–11.2
Total COD (g/L)	18.9–20.2
Soluble COD (g/L)	0.5–1.2

2.2. Pre and Post-Treatment Conditions

ULS treatment was performed with an ultrasonicator (Misonix, Q700, Qsonica, CT, USA) at 20 kHz. The power rating of the ultrasonicator is 700 W. During ultrasonication the temperature was monitored and maintained at about 30 °C with an ice-water bath. According to the results of previous studies [19], the specific energy input was selected at 9 kJ/g TS. ULS-Ozone treatment was conducted by applying ozonation after the ULS treatment. The ozonation was performed with an ozone generator (Wedeco, GSO 30, Xylem Water Solutions Herford GmBH, Herford, Germany). A stone diffuser was installed to produce fine ozone bubbles and to enhance ozone mass transfer. The applied ozone dosage of 0.012 g O_3 g^{-1} TS was selected based on previous results [20,21]. ULS+ALK treatment was applied by ultrasonicating the sludge which was being mixed at 200 rpm at a NaOH (Sigma-Aldrich, St Louis, MO, USA) concentration of 0.02 M according to previous research [20,22]. The NaOH concentration was reached by adding a 3 M stock solution into the sludge. The ULS+ALK post-treated digested sludge was neutralized with 6 M HCl before being recycled back to the anaerobic digester.

2.3. Anaerobic Digestion Tests

Anaerobic digestion was conducted semi-continuously in 1.2 L continuously-stirred glass bottles (one control and three test reactors with different post-treatment) with 1 L working volume at 35 °C (Figure 1). Three test reactors included an ULS post-treatment (ULS reactor), ULS-Ozone post-treatment (ULS-Ozone reactor), and ULS+ALK post-treatment (ULS+ALK reactor). Each reactor was seeded with 1 L anaerobic inoculum which was taken from a continuous full-scale anaerobic digester operating at 28 days HRT. Each reactor was then fed with untreated sludge. Biogas produced from each reactor was measured daily with a gas meter (Ritter Apparatebau Gmbh, Bochum, Germany).

Before starting the post-treatment tests, all reactors were operated at 10 days HRT for 20 days to stabilize the reactor and obtain similar reactor performance. Afterwards, a specific amount of sludge was treated and the recycle ratio R was calculated as follows: R = Q_R/Q_{in} (%). Three different conditions were tested in the reactors as shown in Table 2: Condition I: HRT = 10 days and R = 50%; Condition II: HRT = 10 days and R = 100%; and Condition III: HRT = 20 days and R = 100%. Feeding, withdrawal, and recycling of sludge was conducted manually once a day. The recycle ratio (R) was calculated as the ratio of recycled sludge (Q_R) to the influent flowrate (Q_{in}). For a recycle ratio of 100%, the same volume of fresh sludge and post-treated sludge are added to the reactor, so the reactor receives half its feed as fresh sludge. The post-treatment factor (α) was calculated as the ratio of daily recycled sludge volume to the reactor working volume as defined by Li et al. [12].

Table 2. Operational conditions of each reactor.

Operational Conditions	Condition I	Condition II	Condition III
Duration of the Experiment (days)	15	15	31
HRT = V/Q_{in} (days)	10	10	20
Influent Flowrate, Q_{in} (mL/d)	100	100	50
Recycle Ratio, R = Q_R/Q_{in} (%)	50	100	100
Post-Treatment Factor, α = Q_R/V (%)	5	10	5

2.4. Analytical Methods

COD and solids concentrations were measured in accordance with standard methods [23]. Soluble and total COD were measured based on the closed reflux colorimetric method. The soluble samples were obtained by first centrifuging the sludge at 10,000 rpm for 10 min. The supernatant was then filtrated through 0.45 μm syringe filters for soluble COD analysis. Sludge dewaterability was measured with capillary suction time (CST) as described in standard methods [23]. Sludge pH was measured with a pH meter (Agilent, model 3200P, Santa Clara, CA, USA). VFAs concentration was analyzed with an Agilent Gas chromatograph (Agilent Technologies 7890A GC system, Santa Clara, CA, USA) with a flame ionization detector. The composition of biogas was measured with gas chromatography (Agilent Technologies 7890A GC system, Santa Clara, CA, USA) with a thermal conductivity detector [19].

Adenosine tri-phosphate (ATP) concentration was measured immediately after sampling using QuenchGone21™ Wastewater Test Kit following the manufacturer's instructions (LuminUltra, Fredericton, Canada). The assay is based on the conversion of chemical energy during luciferase reaction into light energy. The emitted light was quantified using a luminometer in relative light units (RLUs) which were converted to actual ATP concentrations (ng/mL) after calibration with 1 ng/mL standard. Cellular and dissolved ATP were measured or calculated according to a procedure explained elsewhere [24]. Additionally, the biomass stress index (BSI) was calculated as the ratio of dead-cell ATP to total ATP [24].

2.5. Statistical Analysis

The results are presented as mean ± standard deviation (SD) together with the number of data points. T-tests to determine statistical differences between treatments were carried out by comparing the critical value through ANOVA one-way analysis of variance (SPSS Statistics V17.0, IBM, New York, NY, USA). Comparisons were considered significantly different at $p < 0.05$.

3. Results

3.1. Biogas Production

The daily biogas production from each reactor is shown in Figure 2. The daily gas production from the four reactors were similar in the first 20 days stabilization period, indicating the performance of each reactor was similar before the post-treatment was applied. Incorporation of the post-treatment improved the daily biogas production from 20 days onwards.

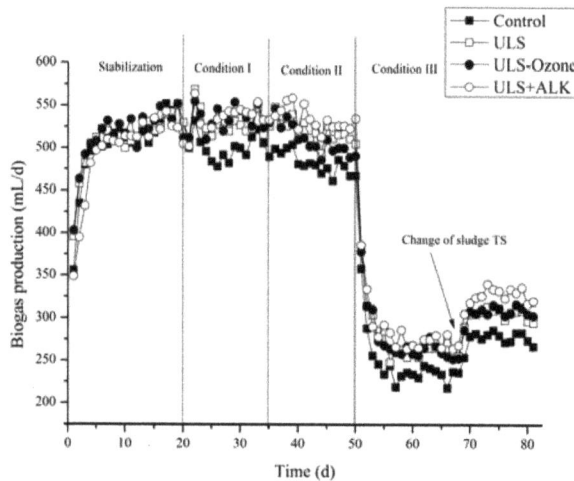

Figure 2. Daily biogas production from the control and test reactors. A fraction of the digested sludge was treated by ultrasound (ULS), ultrasound and ozone (ULS-Ozone) or ultrasound and alkali (ULS+ALK). Condition I: HRT = 10 days and R =50%; Condition II: HRT = 10 days and R = 100%; and Condition III: HRT = 20 days and R = 100%.

The methane composition in biogas was around 64% in all the tests, indicating post-treatments did not affect the methane composition. In Condition I, the biogas production due to the ULS, ULS-Ozone, and ULS+ALK post-treatment were, respectively, 5.2%, 7.1%, and 8.2% greater than in the control. This was achieved at R = 50% meaning that 50 mL/day of digested sludge going though post-treatment is mixed with 100 mL/day of raw feed. This post-treatment configuration would, therefore, consume half the energy required for the corresponding pre-treatment configuration (100 mL/day would have to be treated through any pre-treatment) while still achieving a significant biogas increase.

In Condition II, The ULS, ULS-Ozone, and ULS+ALK post-treatments increased the daily biogas production by 8%, 4.9%, and 11.1%, respectively. The biogas production due to the ULS (8%) and ULS+ALK (11.1%) post-treatments were higher than in condition I (5.2% and 8.2%, respectively). This is because more digested sludge was post-treated and recycled as substrate (higher R). However, the biogas production due to the ULS-Ozone post-treatment decreased when the volume of recycled sludge doubled. Furthermore, the T-test confirmed that the daily biogas production from the ULS-Ozone reactor was statistically lower than that produced from the ULS reactor. Li et al. [12] observed a decrease in biogas production when the recycled sludge (treated with 0.1M NaOH for 30 min) exceeded 5% of the total working volume of the anaerobic digester ($\alpha > 5\%$). According to Li et al. [12], the decrease in biogas production was related to the inactivation of anaerobic microorganisms at higher α values. However, in this study, increase of α from 5% to 10% only decreased the biogas production from the ULS-Ozone reactor, but increased the biogas production from the ULS and ULS+ALK reactors. Therefore, the influence of α on the biogas production was dependent on the selected treatment methods. In addition, no volatile fatty acids (VFAs) were detected in the effluent of all the reactors during conditions I and II, suggesting that the methanogenesis step was not inhibited even at 10 days HRT.

In Condition III, the biogas increases due to the ULS, ULS-Ozone and ULS+ALK post-treatments were 9.8%, 10.7%, and 17.8%. These increases were statistically greater than the corresponding increases observed in Conditions I and II. This is due to the higher HRT of 20 days applied during Condition III. On one hand, the higher HRT provided more time for the biodegradation of the feed and post-treated sludge. On the other hand, the digested sludge contained less biodegradable solids. The post-treatment energy could solubilize more slowly biodegradable solids, which also benefited the overall anaerobic digestion. Future studies should focus on two-stage anaerobic digestion with an inter-stage physico-chemical treatment.

Previous studies on batch anaerobic digestion assays showed that the ULS-Ozone post-treatment resulted in higher ultimate methane production than the ULS+ALK post-treatment [19–22]. However, this was not the case when the post-treatment was applied in semi-continuous reactors. This is because post-treated digested sludge acted as a substrate and was given sufficient time (30 days) for the degradation during batch assay. In contrast, the hydraulic residence time (HRT) was much shorter in semi-continuous reactors, as shown in Table 2. It is known that addition of ozone unavoidably increased the oxidation-reduction potential of the reactor and may have induced a lag phase. This shortened the degradation time under strict anaerobic conditions in one cycle and might have decreased the biogas recovery rate.

Effluent soluble chemical oxygen demand (SCOD) increased due to the incorporation of the post-treatment and this was accompanied by higher capillary suction times (CST). These recalcitrant organics were the result of the post-treatment that solubilized some non-biodegradable biopolymers when lyzing the anaerobic microorganisms in digested sludge. In addition, humic acid-like substances are also formed as by-products during the anaerobic digestion of the solubilized macromolecules which contributed to the effluent SCOD concentration [25]. Dewaterability of the digested sludge also deteriorated after post-treatment as shown in Table 3. This was related to the soluble residual biopolymers in the effluent that keeps the solids from being dewatered. The ULS post-treatment was mainly responsible for the increase in effluent SCOD and CST. The combination of alkaline and ULS treatment did not make the effluent SCOD and dewaterability worse. This was in accordance with a previous work by Li and co-workers [12] where individual alkaline post-treatment (0.1 mol/L NaOH) had negligible impacts on the SCOD and dewaterability in the digested sludge.

Table 3. Summary of anaerobic reactors performance when a post treatment was applied to digested sludge. ULS: ultrasound post-treatment; ULS-Ozone: ultrasound and ozone post-treatment; ULS+ALK: ultrasound and alkali post-treatment.

Performance Parameter	Control	ULS	ULS-Ozone	ULS+ALK
Condition I: 10 days HRT, RR = 50%, α = 5%				
Daily Biogas Production (mL/d) (n = 9)	500 ± 12	526 ± 9	525 ± 12	541 ± 6
Methane Yield (mL CH_4/g VS_{added}) (n = 9)	256 ± 5	269 ± 5	268 ± 6	277 ± 3
Effluent TS (mg/L) (n = 5)	11,460 ± 481	10,720 ± 309	10,830 ± 292	11,390 ± 392
Effluent TSS (mg/L) (n = 5)	9980 ± 220	9710 ± 606	9240 ± 487	9840 ± 198
Effluent VS (mg/L) (n = 5)	8470 ± 333	7940 ± 420	8160 ± 429	8140 ± 397
Effluent VSS (mg/L) (n = 5)	7840 ± 219	7530 ± 202	7280 ± 394	7630 ± 211
SCOD (mg/L) (n = 5)	182 ± 6	224 ± 7	237 ± 8	220 ± 7
CST (s) (n = 3)	64.7 ± 4.3	113.8 ± 8.3	148.7 ± 7.4	128.1 ± 6.8
Condition II: 10 days HRT, RR = 100%, α = 10%				
Daily Biogas Production (mL/d) (n = 9)	474 ± 8	512 ± 9	498 ± 8	526 ± 7
Methane Yield (mL CH_4/g VS_{added}) (n = 9)	249 ± 4	269 ± 5	261 ± 4	276 ± 4
Effluent TS (mg/L) (n = 5)	10,960 ± 378	10,780 ± 275	10,500 ± 252	11,300 ± 362
Effluent TSS (mg/L) (n = 5)	9690 ± 368	9570 ± 529	9140 ± 608	9490 ± 595
Effluent VS (mg/L) (n = 5)	8150 ± 406	7920 ± 431	7710 ± 222	7760 ± 347
Effluent VSS (mg/L) (n = 5)	7740 ± 111	7700 ± 82	7310 ± 342	7390 ± 403
SCOD (mg/L) (n = 5)	184 ± 11	228 ± 3	242 ± 3	234 ± 4
CST (s) (n = 3)	63.1 ± 3.3	127.4 ± 5.4	143.9 ± 6.3	131.2 ± 5.1
Condition III: : 20 days HRT, RR = 100%, α = 5%				
Daily Biogas Production (mL/d) (n = 11)	279 ± 5	306 ± 5	309 ± 5	329 ± 7
Methane Yield (mL CH_4/g VS_{added}) (n = 9)	275 ± 5	301 ± 5	304 ± 5	324 ± 7
Effluent TS (mg/L) (n = 5)	11,820 ± 480	11,320 ± 649	11,470 ± 160	12,230 ± 850
Effluent TSS (mg/L) (n = 5)	10,560 ± 227	9850 ± 173	9800 ± 509	9860 ± 403
Effluent VS (mg/L) (n = 5)	8710 ± 399	8160 ± 282	8080 ± 354	8210 ± 530
Effluent VSS (mg/L) (n = 5)	8460 ± 393	7750 ± 364	7540 ± 531	7700 ± 285
SCOD (mg/L) (n = 5)	225 ± 6	245 ± 4	270 ± 11	246 ± 4
CST (s) (n = 3)	74.6 ± 4.4	134.8 ± 5.7	156 ± 5	143.8 ± 5.4

TS: total solids; TSS: total suspended solids; VS: volatile solids; VSS: volatile suspended solids; SCOD: soluble chemical oxygen demand; CST: capillary suction time; n: number of days at the end of the experiment during which the data were averaged.

It should be noted that the digested sludge from the ULS-Ozone reactor had slightly higher SCOD and CST compared to the digested sludge from the ULS reactor. Application of ozonation subsequent to the ULS post-treatment increased the effluent SCOD concentration from 224 to 237 mg/L in condition I. However, such increase was statistically insignificant compared to the change caused by the ULS post-treatment (from 182 to 224 mg/L). The increases in biogas production were statistically significant as confirmed by the t-test provided in Table 4. In addition, the t-test results also showed that the biogas increase due to the ULS-Ozone and ULS+ALK post-treatments were statistically higher than ULS alone, showing the chemical methods were adding value to the ultrasound post-treatment.

Table 4. Statistical analysis of the biogas production increase and effluent VS decrease due to post-treatment at different conditions.

Statistical Parameter	Control	ULS	ULS-Ozone	ULS+ALK
Condition I: 10 days HRT, RR = 50%, α = 5%				
Daily Biogas Production Increase (%)	-	5.2	7.1	8.2
T-Test Compared to Control	-	7.83 [a]	7.65 [a]	11.06 [a]
T-Test Compared to ULS	-	-	3.63 [a]	7.67 [a]
Decrease in Effluent VS (%)	-	6.3	3.7	3.9
T-Test Compared to Control	-	−4.17 [b]	−2.48 [c]	−2.8 [b]
T-Test Compared to ULS	-	-	1.09 [c]	0.83 [c]
Condition II: 10 days HRT, RR = 100%, α = 10%				
Daily Biogas Production Increase (%)	-	8	4.9	11.1
T-Test Compared to Control	-	12.67 [a]	9.44 [a]	16.75 [a]
T-Test Compared to ULS	-	-	−10.93 [b]	5.8 [a]
Decrease in Effluent VS (%)	-	2.8	5.4	4.8
T-Test Compared to Control	-	−1.76 [c]	−3.96 [b]	−1.77 [c]
T-Test Compared to ULS	-	-	−2.09 [c]	−0.61 [c]
Condition III: 20 days HRT, RR = 100%, α = 5%				
Daily Biogas Production Increase (%)	-	9.8	10.7	17.8
T-Test Compared to Control	-	22.6 [a]	29.6 [a]	24.21 [a]
T-Test Compared to ULS	-	-	1.84 [c]	16.68 [a]
Decrease in Effluent VS (%)	-	6.3	7.2	5.7
T-Test Compared to Control	-	−5.5 [b]	−6.68 [b]	−1.96 [c]
T-Test Compared to ULS	-	-	−1.21 [c]	0.21 [c]

[a] Significant higher (p-value larger than 2.306); [b] Significant lower (p-value smaller than −2.306); [c] Not significant higher (p-value between −2.306 and 2.306). VS: volatile solids.

3.2. Microbial Stresss during Semi-Continuous Anaerobic Digestion

The ATP distribution in digested sludge is shown in Figure 3. All the post-treatments under all HRTs and the recycle ratio tested resulted in a lower cellular ATP compared to the control despite the small proportion of post-treated sludge compared to the total reactor volume (small α values). The effect on dissolved ATP was marginal. The decrease in the cellular ATP concentration indicated the decrease in microorganisms' activity due to the post-treatment which was not shown in earlier studies. Interestingly, this lower cellular ATP did not prevent higher biogas production when a post-treatment was applied.

The BSI, the ratio between the dissolved and total ATP concentrations, was used to quantify the stress of microbial communities in the anaerobic reactor. Surprisingly, the BSI in the ULS-Ozone reactor was slightly lower than in the ULS reactor. This meant that the use of ozone in the post-treatment did not impose further stress on the anaerobic reactor compared to ULS alone.

However, the BSI in the ULS+ALK reactor was the highest under all conditions tested. This can be due to the accumulation of dissolved solids (i.e., sodium ions). In addition, the increase in BSI was more obvious in conditions I and II (10 days HRT). For example, the BSI increased from 40.1 to 55.7% in

Condition I, but only increased from 28.4 to 32.4% in Condition III. This would suggest the ULS+ALK post-treatment imposed more stress to the anaerobic reactor when HRT was 10 days. Furthermore, when looking at all the BSI during the three consecutive test periods (I to III), it was found that the stress gradually decreased which could be due to an acclimatization and adaptation of microbial communities to the corresponding post-treatment over time. Future studies should, therefore, look at the long-term performance of such post-treatments.

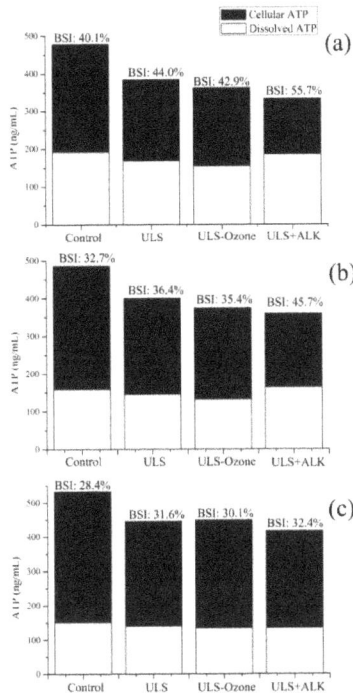

Figure 3. ATP distribution of each reactor at Condition I (**a**); Condition II (**b**); and Condition III (**c**).

4. Discussion

The post-treatment also had impacts on the characteristics of the digested sludge as shown in Table 3. The solids concentration (i.e., TS, TSS, VS, and VSS) in the digested sludge were determined by averaging the corresponding concentrations from five sampling days. A t-test was conducted to statistically compare the results, as shown in Table 4. In many cases, the average effluent VS concentrations from the ULS, ULS-Ozone, and ULS+ALK reactors were lower, but not statistically significant compared to the effluent VS concentration from the control reactor. The post-treatment could obviously improve the biogas recovery from sludge anaerobic digestion while its effects on VS destruction were relatively limited (in the range 4–7%). This was because the recycling of the post-treated sludge increased the biodegradable organic loadings of the reactor which benefited the biogas production.

The tested post-treatments showed the highest VS removals in Condition III. This was because longer residence time was given for the hydrolysis of feed and post-treated sludge. Battimelli et al. [14] showed that the COD and solids removal rates started to decrease when the recycle ratio (R) between recycled sludge (treated with 0.16 g O_3/g TS) to feed sludge exceeded 25% due to the reduction of SRT. This reduction is caused by cell lysis in the recycle line due to the post-treatment. These results confirm the importance of an appropriate recycle ratio and sufficient residence time of the anaerobic reactor with the post-treatment incorporated.

The t-test results also indicated neither the ULS-Ozone nor ULS+ALK post-treatment showed obvious increases in VS removal compared to ULS post-treatment in the tested conditions, indicating the chemical methods did not significantly benefit the solids removal caused by the ULS treatment. Although effluent TS concentration decreased for ULS and ULS-Ozone reactors as a result of the VS destruction, the effluent TS was similar to the control reactor during Condition I. However, effluent TS was slightly higher for the ULS+ALK reactor (from 11.82 g/L in the control to 12.23 g/L in Condition III) due to NaOH addition which increased the dissolved solids concentration in the reactor over time. This increase is consistent with our NaOH dosage of 0.02 M or 800 mg/L. This is in contrast with the literature that reported a decrease in TS due to ULS+ALK pre-treatment in batch mode: in Seng et al. [26], the TS removal increased from 12.5% (control digester) to 17% with a chemical dose of 15 mg g^{-1} TS and then continued increasing to around 18% when the chemical dose increased to 25 mg g^{-1} TS. However, in continuous reactor with 10 mg NaOH/g TS, the TS removal was only 2% at 25 days HRT. The authors explained that the low TS removal for chemical–ultrasound pretreated WAS was due to the addition of NaOH, which contributed to the TS content. This work confirmed that low TS removals can be expected when ULS+ALK was used as post-treatment.

Li et al. [12] indicated the potential risk at α factor of 10% and 15% while the reactor was operated at 20 days HRT. This emphasized the importance of choosing an appropriate recycle ratio, especially when the alkaline treatment is applied. Although inhibition due to dissolved solids (e.g., sodium ions) was not observed in Li et al. [12] or this study, the risk of sodium inhibition is present over time. Moreover, the contamination of excess sludge with sodium may require special disposal considerations.

The biogas production increase in the semi-continuous anaerobic digestion reactors due to pre- and post-treatments are compared in Table 5. The ULS-Ozone pre-treatment resulted in higher biogas production increase than the ULS-Ozone post-treatment at 10 and 20 days HRT. This indicated the ULS-Ozone was more suitable for treatment of feed sewage sludge than for treatment of digested sludge in enhancing biogas production. This is related to the effects of the treated sludge on the anaerobic digestion process. The feed sludge acted as substrate for anaerobic digestion; whereas, the digested sludge not only acted as substrate for the anaerobic digestion, but also contained active anaerobic microorganisms which were essential for the anaerobic digestion [12,14]. Consequently, the post-treatment method can have negative effects such as the inactivation of anaerobic bacteria in digested sludge. Therefore, the lower biogas production increase observed in the ULS-Ozone post-treatment configuration was due to the inactivation or lysis of essential anaerobic microorganisms (e.g., hydrogenotrophic methanogens) in the digested sludge. This could have negated its positive effects on the biodegradability improvement.

Similarly, the pre-treatment configuration was more advantageous than the post-treatment configuration in terms of enhancing biogas production at HRT of 10 days for the ULS and ULS+ALK treatments. In contrast, the post-treatment configuration performed slightly better at HRT of 20 days for these treatments.

Full-scale application of ultrasound to pre-treat sludge was reported to result in 13–58% increase in biogas and up to 22% solids destruction at an energy input of 1.44 kWh/m^3 of treated sludge [27]. A small laboratory scale probe was used in this study which required a significantly higher energy input of 9000 kJ/g TS (equivalent to 25 kWh/m^3) to observe similar performance. Nevertheless, the combination of ozone or ALK with ultrasound are unlikely to justify the additional energy demand given the increment in biogas production compared to ULS alone. Based on the laboratory data, the application of ULS post-treatment seems justified and future studies could investigate the inclusion of a ULS step in-between two digesters.

Table 5. Comparison of performance of pre-treatment and post-treatment using ULS, ULS-Ozone, and ULS+ALK treatments at 10 and 20 days HRT.

Performance Parameter	Treatment (HRT, R)	ULS	ULS-Ozone	ULS+ALK
Biogas Increase (%)	Pre-treatment (10)	20.7	35.9	24.6
	Pre-treatment (20)	7.7	25.5	16.6
	Post-treatment (10, 50%)	5.2	7.1	8.2
	Post-treatment (10, 100%)	8	4.9	11.1
	Post-treatment (20, 100%)	9.8	10.7	17.8
Solids Removal (%)	Pre-treatment (10)	7.6	18.3	15.7
	Pre-treatment (20)	9.7	21.4	18.2
	Post-treatment (10, 50%)	11.7	6.8	7.3
	Post-treatment (10, 100%)	4.7	9.1	8
	Post-treatment (20, 100%)	9.5	10.9	8.6
Post-Digestion SCOD Concentration (mg/L)	Pre-treatment (10)	194 to 257	194 to 589	194 to 296
	Pre-treatment (20)	182 to 227	182 to 440	182 to 246
	Post-treatment (10, 50%)	182 to 224	182 to 236	182 to 220
	Post-treatment (10, 100%)	184 to 228	184 to 242	184 to 234
	Post-treatment (20, 100%)	225 to 245	225 to 270	225 to 246

HRT: hydraulic retention time (days); R: recycle ratio in post-treatment (%).

In terms of solids removal, ULS-ozone and ULS-ALK achieved better results in pre-treatment configuration regardless of the HRT and recycle ratio. However, The ULS post-treatment at 10 days HRT and 50% recycle ratio achieved better removal than in pre-treatment (11.7% versus 7.6%). Moreover, this was achieved with only 50 mL of sludge being post-treated, whereas 100 mL was treated in the pre-treatment configuration. This indicated the potential of ULS to be used as post-treatment using half the amount of sludge, hence, half the energy input. At 20 days HRT, both pre- and post-ULS treatment achieved about 9.5% solids removal, which was consistent with the corresponding increase in biogas production. All configuration (pre and post) resulted in an increase in the final effluent SCOD, which translated to an increase in capillary suction times.

5. Conclusions

This work showed that the post-treatments were able to increase the biogas production and decrease the VS in the final effluent. The maximum daily biogas increase was 17.8% when the ULS+ALK post-treatment was applied to a reactor operating at 20 days HRT and 100% recycle ratio. At 50% recycle ratio (Condition I), biogas increase in the range 5–8% can be achieved at half the energy input required in a comparable pre-treatment configuration. Based on the results, the post-treatment of digested sludge or treating the sludge between two digesters is an interesting alternative to pre-treatments.

Acknowledgments: The authors would like to express sincere thanks to the Public Utilities Board (PUB), Singapore for sponsoring the project and providing the sludge and Xylem Water Solutions Herford GmBH for providing the ozone generator.

Author Contributions: Tian Xinbo conceived and designed the experiments; Tian Xinbo performed the experiments; and Tian Xinbo and Antoine Trzcinski analyzed the data and wrote the paper.

References

1. Eastman, J.A.; Ferguson, J.F. Solubilization of particulate organic carbon during the acid phase of anaerobic digestion. *J. Water Pollu. Control Fed.* **1981**, *53*, 352–366.
2. Pavlostathis, S.G.; Giraldo-Gomez, E. Kinetics of Anaerobic Treatment. *Water Sci. Technol.* **1991**, *25*, 35–59.
3. Stuckey, D.C.; McCarty, P.L. Thermochemical pretreatment of nitrogenous materials to increase methane yield. *Biotechnol. Bioeng. Symp.* **1978**, *8*, 219–233.

4. Stuckey, D.C.; McCarty, P.L. The Effect of Thermal Pretreatment on the Anaerobic Biodegradability and Toxicity of Waste Activated Sludge. *Water Res.* **1984**, *18*, 1343–1353. [CrossRef]
5. Tiehm, A.; Nickel, K.; Neis, U. The use of ultrasound to accelerate the anaerobic digestion of sewage sludge. *Water Sci. Technol.* **1997**, *36*, 121–128. [CrossRef]
6. Bougrier, C.; Albasi, C.; Delgenes, J.; Carrere, H. Effect of ultrasonic, thermal and ozone pre-treatments on waste activated sludge solubilisation and anaerobic biodegradability. *Chem. Eng. Process.* **2006**, *45*, 711–718. [CrossRef]
7. Xu, G.; Chen, S.; Shi, J.; Wang, S.; Zhu, G. Combination treatment of ultrasound and ozone for improving solubilization and anaerobic biodegradability of waste activated sludge. *J. Hazard. Mater.* **2010**, *180*, 340–346. [CrossRef] [PubMed]
8. Kim, D.-H.; Jeong, E.; Oh, S.-E.; Shin, H.-S. Combined (alkaline+ultrasonic) pretreatment effect on sewage sludge disintegration. *Water Res.* **2010**, *44*, 3093–3100. [CrossRef] [PubMed]
9. Gossett, J.M.; Stuckey, D.C.; Owen, W.F.; McCarty, P.L. Heat treatment and anaerobic digestion of refuse. *J. Environ. Eng. Div.* **1982**, *108*, 437–454.
10. Takashima, M.; Kudoh, Y.; Tabata, N. Complete anaerobic digestion of activated sludge by combining membrane separation and alkaline heat post-treatment. *Water Sci. Technol.* **1996**, *34*, 477–481. [CrossRef]
11. Nielsen, H.B.; Thygesen, A.; Thomsen, A.B.; Schmidt, J.E. Anaerobic digestion of waste activated sludge—Comparison of thermal pretreatments with thermal inter-stage treatments. *J. Chem. Technol. Biotechnol.* **2010**, *86*, 238–245. [CrossRef]
12. Li, H.; Zou, S.; Li, C.; Jin, Y. Alkaline post-treatment for improved sludge anaerobic digestion. *Bioresour. Technol.* **2013**, *140*, 187–191. [CrossRef] [PubMed]
13. Takashima, M. Examination on Process Configurations Incorporating Thermal Treatment for Anaerobic Digestion of Sewage Sludge. *J. Environ. Eng.* **2008**, *134*, 543–549. [CrossRef]
14. Battimelli, A.; Millet, C.; Delgenesm, J.P.; Moletta, R. Anaerobic digestion of waste activated sludge combined with ozone post-treatment and recycling. *Water Sci. Technol.* **2003**, *48*, 61–68. [PubMed]
15. Goel, R.; Yasui, H.; Shibayama, C. High-performance closed loop anaerobic digestion using pre/post sludge ozonation. *Water Sci. Technol.* **2003**, *47*, 261–267. [PubMed]
16. Rivero, J.A.C.; Madhavan, N.; Suidan, M.T.; Ginestet, P.; Audic, J.-M. Enhancement of anaerobic digestion of excess municipal sludge with thermal and/or oxidative treatment. *J. Environ. Eng.* **2006**, *132*, 638–644. [CrossRef]
17. Takashima, M.; Tanaka, Y. Application of acidic thermal treatment for one- and two-stage anaerobic digestion of sewage sludge. *Water Sci. Technol.* **2010**, *62*, 2647–2654. [CrossRef] [PubMed]
18. Takashima, M.; Tanaka, Y. Acidic thermal post-treatment for enhancing anaerobic digestionof sewage sludge. *J. Environ. Chem. Eng.* **2014**, *2*, 773–779. [CrossRef]
19. Tian, X.; Wang, C.; Trzcinski, A.P.; Lin, L.; Ng, W.J. Interpreting the synergistic effect in combined ultrasonication–ozonation sewage sludge pre-treatment. *Chemosphere* **2015**, *140*, 63–71. [CrossRef] [PubMed]
20. Tian, X.; Trzcinski, A.P.; Lin, L.L.; Ng, W.J. Enhancing sewage sludge anaerobic "re-digestion" with combinations of ultrasonic, ozone and alkaline treatments. *J. Environ. Chem. Eng.* **2016**, *4*, 4801–4807. [CrossRef]
21. Tian, X.; Trzcinski, A.P.; Lin, L.L.; Ng, W.J. Impact of ozone assisted ultrasonication pre-treatment on anaerobic digestibility of sewage sludge. *J. Environ. Sci.* **2015**, *33*, 29–38. [CrossRef] [PubMed]
22. Tian, X.; Wang, C.; Trzcinski, A.P.; Lin, L.; Ng, W.J. Insights on the solubilization products after combined alkaline and ultrasonic pre-treatment of sewage sludge. *J. Environ. Sci.* **2015**, *29*, 97–105. [CrossRef] [PubMed]
23. American Public Health Association (APHA). *Standard Methods for the Examination of Water and Wastewater*, 22th ed.; APHA: Washington, DC, USA, 2012.
24. Trzcinski, A.P.; Ganda, L.; Kunacheva, C.; Zhang, D.Q.; Lin, L.L.; Tao, G.; Lee, Y.; Ng, W.J. Characterization and biodegradability of sludge from a high rate A-stage contact tank and B-stage membrane bioreactor of a pilot-scale AB system treating municipal wastewaters. *Water Sci. Technol.* **2016**, *74*, 1716–1725. [CrossRef] [PubMed]
25. Luo, K.; Yang, Q.; Li, X.-M.; Chen, H.-B.; Liu, X.; Yang, G.-J.; Zeng, G.-M. Novel insights into enzymatic-enhanced anaerobic digestion of waste activated sludge by three-dimensional excitation and emission matrix fluorescence spectroscopy. *Chemosphere* **2013**, *91*, 579–585. [CrossRef] [PubMed]

26. Seng, B.; Khanal, S.K.; Visvanathan, C. Anaerobic digestion of waste activated sludge pretreated by a combined ultrasound and chemical process. *Environ. Technol.* **2010**, *31*, 257–265. [CrossRef] [PubMed]
27. Xie, R.; Xing, Y.; Abdul Ghani, Y.; Ooi, K.-E.; Ng, S.-W. Full-scale demonstration of an ultrasonic disintegration technology in enhancing anaerobic digestion of mixed primary and thickened secondary sewage sludge. *J. Environ. Eng. Sci.* **2007**, *6*, 533–541. [CrossRef]

environments

MDPI

Article

Removal of Synthetic Dyes by Dried Biomass of Freshwater Moss *Vesicularia Dubyana*: A Batch Biosorption Study

Martin Pipíška [1,2], Martin Valica [1], Denisa Partelová [1], Miroslav Horník [1,*], Juraj Lesný [1] and Stanislav Hostin [1]

[1] Department of Ecochemistry and Radioecology, University of SS. Cyril and Methodius in Trnava, Nam. J. Herdu 2, 91701 Trnava, Slovakia; pipiskam@gmail.com (M.P.); martin.valica@outlook.sk (M.V.); d.partelova@gmail.com (D.P.); juraj.lesny@ucm.sk (J.L.); stanislav.hostin@ucm.sk (S.H.)
[2] Department of Chemistry, Trnava University in Trnava, Priemyselná 4, 91843 Trnava, Slovakia
* Correspondence: miroslav.hornik@ucm.sk; Tel.: +421-335-565-392

Received: 27 November 2017; Accepted: 6 January 2018; Published: 9 January 2018

Abstract: In this work the biosorption of cationic dyes thioflavin T (TT) and methylene blue (MB) from single and binary solutions on dried biomass of freshwater moss *Vesicularia dubyana* as a function of contact time, pH, and biomass or sorbate concentration has been investigated. The prediction of maximum sorption capacities using adsorption isotherm models were also realized. Biosorption of TT and MB is a rapid process strongly affected by solution pH. Maximum sorption capacities Q_{max} calculated from Langmuir isotherm were 119 ± 11 mg/g for TT and 229 ± 9 mg/g for MB. In binary mixture, the presence of MB caused significant decrease of TT sorption, advocating the competitive sorption between TT and MB. Results revealed that *V. dubyana* biomass exhibited significantly higher affinity to thiazine dye MB in comparison with benzothiazole dye TT from both single and binary solutions. Based on the obtained results, the competitive effects in binary system can substantially influence the sorption process and should be thoroughly evaluated before application of selected adsorbents for removal of basic dyes from colored effluents.

Keywords: biosorption; cationic dyes; moss; *Vesicularia dubyana*

1. Introduction

Dyes are the important class of synthetic organic compounds used in many industries such as manufacture of pulp and paper, leather tanning or textile dying and in manufacture of dyestuffs. Ghaly et al. [1] pointed out that the textile industry is one of the major industries in the world and it plays a major role in the economy of many countries. The textile industry utilizes various chemicals (particularly synthetic dyes) and large amounts of water during the production process. The global textile dyes market was estimated to reach USD 4.7 billion in 2015, and is projected to reach USD 6.4 billion by 2019 and USD 8.75 billion by 2023 [2,3].

Industrial application of synthetic dyes is associated with the release of a huge amount of more or less colored effluents into the environment [4,5]. Wastewaters from the textile industry contain a large amount of dyes and chemicals containing trace metals such as Cr, As, Cu and Zn which are capable of harming the environment and human health [1]. During textile dyeing, a wide range of various dyes in a short time period are used, therefore effluents are extremely variable in composition and require an unspecific treatment processes [5]. Conventional physico-chemical (dilution, adsorption, coagulation and flocculation, oxidation, reverse osmosis and ultrafiltration) and biological (aerobic activated sludge and anaerobic processes) treatment technologies presently employed for color removal have several disadvantages such as long operational time, low specificity, formation and disposal of sludge and high cost (see e.g., critical reviews [1,6,7]. Consequently, the development of efficient clean-up technologies is of major interest.

Highly efficient removal of synthetic dyes from effluents has been attempted by bioadsorption and non-conventional biosorbents have been searched in recent years that are easy available, renewable and environmentally friendly that can successfully replace the classical adsorbents [8]. To date, diverse algae, macroalgae, plant biomasses, agricultural residues as well as other biomaterials have been explored as adsorbents for removal of acidic, basic and reactive dyes from aqueous solutions [9–17]. Authors have reported that the biosorption capacity is highly dependent on solution pH, biosorbent dosage, temperature and concentration of other solutes present in solution. Since industrial effluents may contain several synthetic dyes, it is necessary to study the simultaneous sorption of two or more dyes and to quantify the mutual effect of one dye on the other. However, contradictory findings related to biosorption of dyes in multi-dye sorption systems were recently published. Albadarin et al. [18] who studied the simultaneous sorption of methylene blue and alizarin red S by olive stone biomass revealed only the limited competition between dyes. Remenárová et al. [19] confirmed that thioflavin T (TT) significantly affected biosorption of malachite green (MG) by moss *R. squarrosus* in binary system TT + MG. The competitive effect of MG on TT was less pronounced. On the contrary, Giwa et al. [20] found that the presence of rhodamine B and methylene blue had a synergetic effect on the maximum monolayer capacity of the sawdust of *Parkia biglobosa* for Acid Blue 161 dye in binary and ternary systems.

To ensure the applicability of biosorption technology for colored effluents, more works are still needed for the sorption of a mixture of dyes at various operating conditions. Considering the above mentioned aspects, this study concerns the biosorption of cationic dyes thioflavin T and methylene blue from aqueous solutions by dried biomass of freshwater moss *Vesicularia dubyana*. The influence of variables (contact time, pH, biomass dosage) controlling the biosorption process has been considered in both single and binary dye systems.

2. Materials and Methods

2.1. Moss Biomass

The biomass of freshwater moss *Vesicularia dubyana* was used as a sorbent of cationic dyes thioflavin T (TT) and methylene blue (MB). Vital moss biomass was cultivated in diluted Hoagland medium [21] under artificial illumination (2000 lx) at $22 \pm 2\,°C$. Before use in sorption experiments, the biomass was thoroughly rinsed in deionized water (3 times) and dried at $60\,°C$ for 48 h.

2.2. Reagents and Instruments

Thioflavin T (C.I. 49005, $M_r(C_{17}H_{19}ClN_2S)$ 318.86, CAS 2390-54-7) and methylene blue (C.I. 52015, M_r ($C_{16}H_{18}N_3SCl$) 319.86; CAS 61-73-4) were purchased from Fluka (USA). All chemicals were of analytical grades. Cationic dyes solutions were prepared in deionized water (conductivity 0.054 μS/cm; Millipore Simplicity). An UV-VIS spectrophotometer Cary 50 (Varian, Australia) was used for determination of TT and MB concentrations in solutions and for establishing calibration curves for TT and MB at maximum absorbances $\lambda TT = 412$ nm and $\lambda MB = 650$ nm. The pH effect on MB and TT concentration determination was taken into account.

2.3. Sorption Kinetics in Single and Binary Solutions in Batch System

Dried biomass of *V. dubyana* (0.5 g/L) was added to 20 mL of solutions containing defined concentrations of thioflavin T (TT), methylene blue (MB) or their mixtures, respectively. Solution pH was adjusted to 6.0. The flasks were incubated on a rotary shaker (250 rpm) at $25\,°C$. At the time intervals 10, 20, 40, 60, 120, 240, 360 and 1440 min, aliquot samples were obtained and remaining concentration of TT and MB determined. All experiments were performed in duplicate series. The TT and MB uptake was calculated according to Equation (1).

$$Q_t = (C_0 - C_t)\frac{V}{M} \tag{1}$$

where Q_t represents the amount of TT or BB sorbed by moss biomass (mg/g d.w.) from single or binary solutions at time t; C_0 and C_t represent the initial concentration of dyes in solution and the concentration of dyes at time t (mg/g). V is solution volume (L) and M is the amount of moss biomass (g; d.w.).

2.4. Influence of pH and Biomass Dosage

To analyze the influence of pH, dried biomass (0.5 g/L) was shaken in 20 mL of TT and MB ($C_{0\,TT}$ = 40 mg/L or $C_{0\,MB}$ = 80 mg/L) or TT + MB ($C_{0\,TT}$ = 40 mg/L and $C_{0\,MB}$ = 40 mg/L) solutions of desired pH for 2 h on a rotary shaker at 250 rpm and 25 °C. In order to eliminate interference of buffer components on cationic dyes biosorption, the non-buffered solutions in deionized water were adjusted to the desired pH values by adding 0.1 M HCl or 0.1 M NaOH.

To analyze the influence of biomass dosage, dried biomass of desired amount (C_B = 0.25 to 4.0 g/L) was shaken in 20 mL of TT and MB ($C_{0\,TT}$ = 40 mg/L or $C_{0\,MB}$ = 80 mg/L) or TT + MB ($C_{0\,TT}$ = 40 mg/L and $C_{0\,MB}$ = 40 mg/L) solutions for 2 h on a rotary shaker at 250 rpm, pH 6.0 and 25 °C.

At the end of the experiments, aliquot samples were obtained and remaining concentration of TT and MB determined. All experiments were performed in duplicate series. The TT and MB uptake was calculated according to Equation (1).

2.5. Sorption Equilibrium in Single and Binary Solutions in Batch System

Dried biomass of *V. dubyana* (0.5 g/L) was added to 20 mL of solutions with initial dye concentrations C_0 in single system ranging from 20 to 200 mg/L (TT) and 20 to 320 mg/L (MB) and in binary system TT + MB from 40 to 160 mg/L. In both single and binary solutions pH was adjusted to 6.0. Flasks were incubated on a rotary shaker (250 rpm) at 25 °C. After 2 h of exposure, the remaining concentration of TT and MB was determined. All experiments were performed in duplicate series. The TT and MB uptake was calculated according to Equation (1).

Equilibrium data in both single and binary systems were analyzed using adsorption isotherm models according to Langmuir and Freundlich. To calculate the corresponding parameters of isotherms non-linear regression analysis was performed by OriginPro 2016 (OriginLab Corporation, Northampton, MA, USA).

2.6. Potentiometric Titration

Potentiometric titration experiment was carried out according to the modified procedure described by Zhang et al. [22]. Dried biomass of freshwater moss *V. dubyana* (0.30 g) were protonated with 100 mL of equimolar 0.1 mol/L HCl and NaCl for 2 h on the rotary shaker (200 rpm) at 25 °C. Subsequently, moss biomass was separated by centrifugation and transferred into 100 mL of 0.1 mol/L NaCl and agitated under the same conditions. Potentiometric titration was performed in an Erlenmeyer flask with glass electrode (three-point calibrated with buffers—pH 4.0, 7.0 and 10.0) by dropwise addition of 0.1 mol/L mixture of NaOH and NaCl into the bacterial suspension. The titration was conducted in the pH range of 2.0–11.0 and the potentiometric titration curve was obtained by plotting the volume of 0.1 mol/L NaOH addition as titrant solution against the pH values measured. ProtoFit ver. 2.1 was used to analyse the titration data.

3. Results and Discussion

3.1. Biosorption Kinetics in Single and Binary Systems

As expected, the uptake of cationic dyes TT and MB by dried biomass of freshwater moss *V. dubyana* from both single and binary solutions was a rapid process (Figure 1A,B). At the initial phase of uptake, driving force is high (uptake increased linearly) and available high affinity binding sites on *V. dubyana* biomass are occupied. Initial phase is followed by slower gradual uptake till equilibrium is reached.

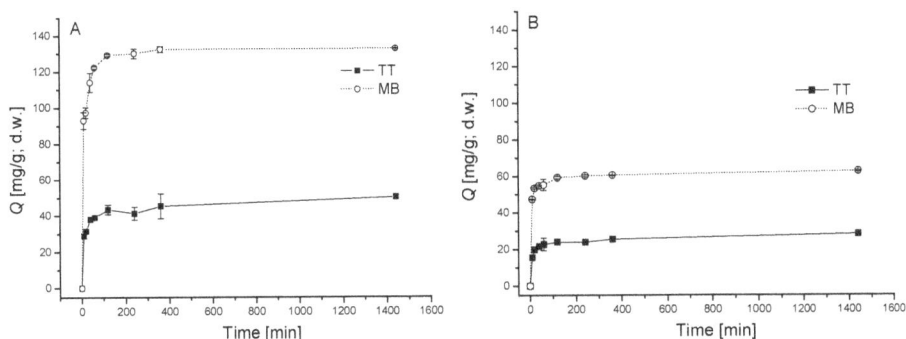

Figure 1. Kinetics of thioflavin T (TT) and methylene blue (MB) biosorption by dried biomass of moss *V. dubyana* (C_B = 0.5 g/L) from single (**A**. $C_{0\ TT}$ = 40 mg/L or $C_{0\ MB}$ =80 mg/L) and binary (**B**. $C_{0\ TT}$ = 40 mg/L and $C_{0\ MB}$ = 40 mg/L) solutions at 25 °C and pH 6.0. Error bars represent standard deviation of the mean ($\pm$SD, n = 2).

In case of MB, the maximum uptake in single system at initial concentration 80 mg/L was observed after 120 min of exposure (130 ± 5 mg/g; d.w.). Uptake of TT at initial concentration 40 mg/L increased rapidly in the first 90 min and after 120 min reached 43.5 ± 2.7 mg/g. The final equilibrium was reached within the 2 h and after this time, there was no considerable increase in both TT and MB sorption until the end of experiments. Our findings are in agreement with other studies of cationic dyes sorption by various types of biomass. Similar behavior of MB sorption kinetics by green alga *Enteromorpha* spp. [23] and red (*Gracilaria parvispora*), brown (*Nizamuddinia zanardinii*) and green (*Ulva fasciata*) macroalga [9] was observed. Hameed [24] found that at higher initial concentration of MB longer contact time is needed to reach equilibrium when grass waste was used as sorbent.

Biosorption of TT and MB by *V. dubyana* biomass from binary solution is shown in Figure 1B. Although the same kinetic profile as in the case of dye sorption in single systems was observed, an evident decrease of TT and MB biosorption capacities in comparison with single systems were recorded as a result of competitive effects (see discussion below). The final equilibrium of both TT and MB was reached within 120 min. At initial dye concentration 40 mg/L, maximum sorption of TT (28.3 ± 0.3 mg/g d.w.) in binary system was significantly lower than maximum sorption of MB (62.7 ± 0.3 mg/g, d.w.). The similar kinetic profile in binary (MB + rhodamine B) and single biosorption system observed Fernandez et al. [25] using *Cupressus sempervirens* cone chips.

It is evident that *V. dubyana* biomass exhibited significantly higher affinity to thiazine dye MB in comparison with benzothiazole dye TT from both single and binary solutions. In both dye molecules the positive charge is present on quaternary nitrogen =N^+=, however in TT quaternary N is located in position 3, substituted by one methyl group and in MB quaternary N is located in position 3 or 7 and substituted by 2 methyl groups. Therefore, we suppose that among other mechanisms, steric effects in dye molecules can affect dye affinity to moss biomass as well.

3.2. Influence of pH

The pH of dye solution is one of the crucial parameters affecting the sorption process through controlling both the ionization of dye molecules and the degree of ionization of functional groups present on biomass surface.

The influence of solution pH (2.0 to 7.0) on TT and MB sorption by *V. dubyana* biomass from single solutions is shown in Figure 2A. The lowest sorption capacities Q for both TT (Q_{TT} = 12 mg/g d.w.) and MB (Q_{MM} = 4 mg/g d.w.) dyes were observed at low initial pH value (pH_0 = 2.0) which could be closely related to protonation of binding sites on biomass surface. The low adsorption capacity of MB can also be explained by the fact that MB is mainly found as undissociated species MB^0 (99% at pH = 2.0 and 86% at pH = 3.0) [26]. Obtained results indicate that an increase in pH has a positive effect on TT and MB sorption, since the competitions between dye cations and protons for the binding sites decreases and such curves (Figure 2A) represent a typical cationic dye sorption behavior [10,18].

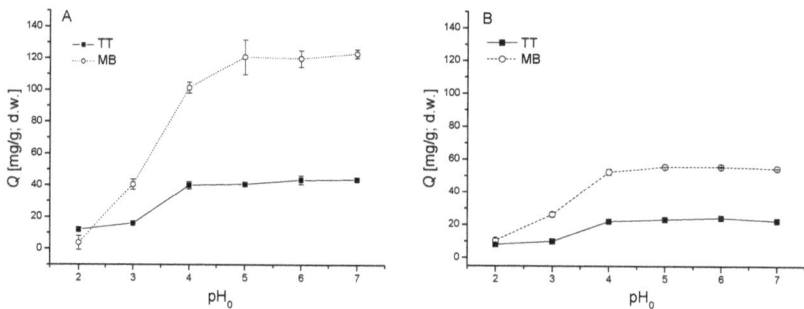

Figure 2. Influence of initial pH_0 on thioflavin T (TT) and methylene blue (MB) biosorption by biomass of moss *V. dubyana* (C_B = 0.5 g/dm^3) from single ($C_{0\ TT}$ = 40 mg/L or $C_{0\ MB}$ 80 mg/L; **A**) and binary ($C_{0\ TT}$ = 40 mg/L and $C_{0\ MB}$ = 40 mg/L; **B**) solutions at 25 °C for 2 h. Error bars represent standard deviation of the mean ($\pm$SD, n = 2).

The highest Q values were recorded at pH > 4.0 with maximum Q_{TT} = 44 $\pm$ 1 mg/g d.w. and Q_{MM} = 123 $\pm$ 3 mg/g d.w. at pH_0 = 7.0. Similarly, Albadarin et al. [23] observed maximum biosorption capacity of methylene blue by olive stone by-products at pH = 7.2 and maximum sorption of thioflavin T by agricultural by-products from the hop (*Humulus lupulus* L.) was found within the range of initial values of pH 4.0–7.0 [27].

The same behavior was observed in binary sorption system with equimolar initial concentrations of TT and MB. Dye sorption capacities increased from 8.2 $\pm$ 0.1 mg/g to 22.7 $\pm$ 0.4 mg/g (TT) and 10.5 $\pm$ 1.4 mg/g to 54.6 $\pm$ 0.1 mg/g (MB) with increasing solution pH from 2.0 to 4.0 and remained almost stable till pH 7.0. From Figure 2B it is evident that the biomass of *V. dubyana* exhibited significantly higher affinity toward MB.

As was mentioned earlier, the dye sorption capacities are to a great extent affected by the dissociation of functional groups present on moss surface. From the potentiometric titration curve (Figure 3A), it is evident that from pH ~4.0 a small addition of NaOH caused a dramatic changes in solution pH what indicates that relevant acidic functional groups (carboxyl, phosphoryl) present on biomass surface are dissociated. To qualitative characterization of functional groups presented on the surface of moss biomass and to determination of binding sites concentration (C_{An}) the prediction modelling was used. The titration curve shows a relatively unpronounced inflection point, which predicts the existence of several functional groups. According to the ProtoFit prediction and obtained values of residual sum of squares (RSS), it was found that the titration profile was best described by a non-electrostatic model for characterization of the four binding sites. The predicted functional groups with relevant *pKa* values and concentrations of binding sites (C_{An}) presented on the surface of moss biomass are listed in Table 1.

In connection to these results, we can suppose that MB$^+$ and TT$^+$ adsorbed on surface and balanced a negative charge of biomass. Consequently, one of the possible mechanism of MB and TT biosorption by freshwater moss *V. dubyana* is the electrostatic attraction.

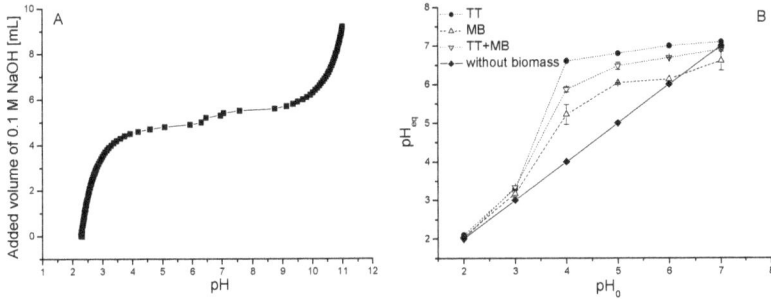

Figure 3. Potentiometric titration curve for dried biomass of freshwater moss *V. dubyana* (3.0 g/L) obtained by gradual addition of titrant 0.1 M NaOH at background electrolyte 0.1 M NaCl and reaction temperature 25 °C (**A**). The change of initial pH$_0$ of dye (single and binary) solutions after biosorption and in control experiments in deionized water and without biomass (**B**).

Table 1. Predicted functional groups according to *pKa* values and concentration of binding sites (C_{An}) presented on the surface of moss biomass (*V. dubyana*) using modelling software ProtoFit ver. 2.1.

Functional Group	pKa	C_{An} (mmol/g)
COOH	2.34–6.25	0.98
PO$_3$H$_2$	6.36–8.72	0.69
NH$_2$	10.7–11.5	0.61
OH	8.36–12.0	0.64

During and after cationic dyes biosorption, the pH of the dye solutions in both single and binary sorption systems changed (Figure 4B). The solution pH increased from initial pH$_0$ values 4.0, 5.0 and 6.0 to pH$_{eq}$ 5.2, 6.0 and 6.1 (MB), 6.6, 6.8 and 7.0 (TT) and 5.8, 6.4, 6.6 (TT + MB). On the contrary, a slight decrease of solution pH was observed from pH$_0$ = 7.0 to pH$_{eq}$ = ~6.6 probably as a result of a TT–hydrogen and/or MB–hydrogen ion-exchange (Figure 3B). This equilibrium value of pH$_{eq}$ also corresponds to the value of pH$_{zpc}$ = 7.78 predicted within the potentiometric titration and ProtoFit analysis, whereby the pH$_{zpc}$ represents the value of pH at which the biosorbent shows a net zero surface charge.

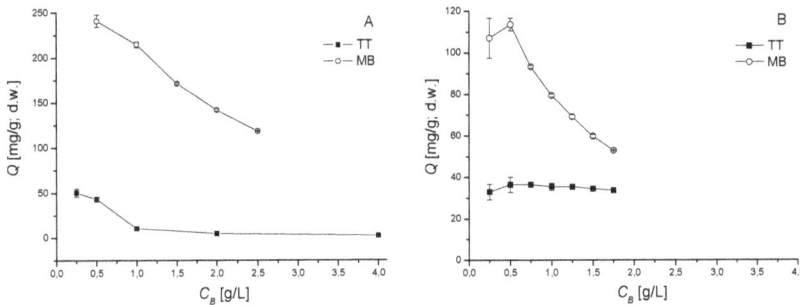

Figure 4. The effect of biomass concentration C_B on thioflavin T (TT) and methylene blue (MB) biosorption by dried biomass of *V. dubyana* from single ($C_{0\ TT}$ = 40 mg/L or $C_{0\ MB}$ = 80 mg/L; (**A**) and binary ($C_{0\ TT}$ = 40 mg/L and $C_{0\ MB}$ = 40 mg/L; (**B**) solutions at 25 °C, pH 6.0 for 2 h. Error bars represent standard deviation of the mean (±SD, *n* = 2).

3.3. Influence of Biomass Dosage

To investigate the biomass dosage on TT and MB removal, various amounts of dried moss biomass were added to both single and binary dye solutions. From Figure 4A it is a clearly seen that the sorption capacity Q of both TT and MB decreased from 241 mg/g to 118 mg/g (MB) and from 50.6 mg/g to 2.6 mg/g (TT) with increasing concentration of moss biomass C_B in single solutions. Different relationship between the sorption capacities of dyes Q and the concentration of biomass C_B was observed in binary sorption system TT + MB (Figure 4B). The sorption capacity of TT changed only slightly with increasing concentration of moss biomass in solution. On the contrary, the sorption capacity Q of MB decreased significantly (from 107 to 52.8 mg/g) with increasing C_B. We suppose that competitive effects between cation dyes and binding sites play an important role and influence the sorption behavior.

Tabaraki and Sadeghinejad [28] pointed out that an increase of biosorbent dose generally increases the overall amount of dye biosorbed (removal efficiency), due to the increased surface area of biosorbent which in turn increases the number of binding sites. Thus, the decrease in the amount of dye sorbed per gram of biosorbent with increase in the biosorbent dose is mainly due to insaturation of binding sites through the sorption process [29]. However, Kumar and Porkodi [30] observed negative aggregation and changes in specific surface area (m^2/g) as well as changes in effective mixing of biomass in sorption systems when higher biomass concentrations were used.

3.4. Sorption Equilibrium in Single and Binary Systems

The adsorption equilibrium data of TT and MB from both single and binary sorption systems were described by adsorption isotherms according to Langmuir and Freundlich.

The isotherm parameters obtained by non-linear regression analysis are reported in Table 2. Coefficients of determination (R^2) related to the Langmuir model applied to TT and MB biosorption data ($R^2 = 0.965$ for TT; $R^2 = 0.990$ for MB) were higher than those for the Freundlich model ($R^2 = 0.955$ for TT; $R^2 = 0.938$ for MB). In addition, the biosorption of methylene blue and thioflavin T by agricultural by-products from the hop (*Humulus lupulus* L.) [27] and methylene blue by water hyacinth biomass [31], agro-waste oil tea shell [13], *Anethum graveolens* biomass [32] and *Tremella fuciformis* biomass [33] was very well fitted by Langmuir model.

Table 2. Langmuir and Freundlich equilibrium parameters ($\pm$SD) obtained by non-linear regression analysis for TT and MB biosorption by *V. dubyana* biomass in single systems.

Dye	Langmuir			Freundlich		
	Q_{max} (mg/g)	b (L/mg)	R^2	K (L/g)	$1/n$	R^2
TT	119 ± 11	0.04 ± 0.01	0.965	15.5 ± 3.8	2.58 ± 0.38	0.955
MB	229 ± 9	0.07 ± 0.01	0.990	45.2 ± 11.5	3.27 ± 0.59	0.938

TT: Thioflavin T; MB: methylene blue.

Maximum sorption capacity Q_{max} and constant b, calculated from Langmuir isotherm model enable to characterize the affinity of *V. dubyana* biomass towards TT and MB (Table 2) in single sorption systems. Q_{max} for TT (119 ± 11 mg/g, d.w.) was significantly lower in comparison with Q_{max} for MB (229 ± 9 mg/g, d.w.). The affinity constant b of the isotherm corresponds to the initial gradient, which indicates the *V. dubyana* biomass affinity at low concentrations of both TT and MB. Accordingly, a higher initial gradient corresponds to a higher affinity constant b. It is evident, that Langmuir isotherm for MB is steeper at lower equilibrium concentrations than those for TT (Figure 5A,B). The difference in the b values 0.04 ± 0.01 L/mg (TT) and 0.07 ± 0.01 L/mg (MB), confirmed the higher affinity of *V. dubyana* biomass to MB in comparison with TT. This was also reflected by higher Q_{max} for MB (Table 2).

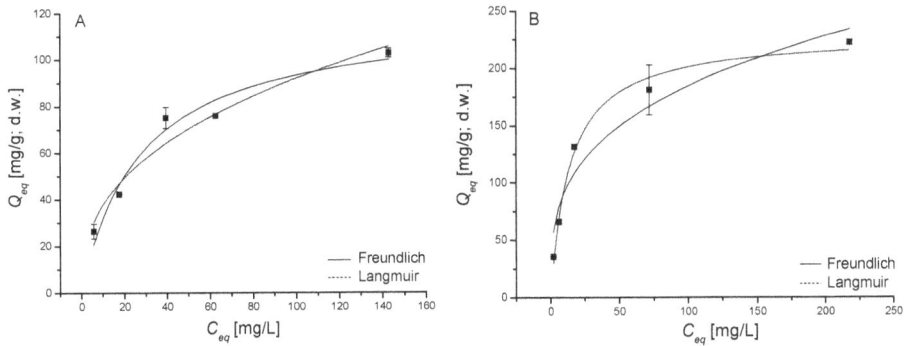

Figure 5. Fit of the Langmuir and Freundlich isotherms of thioflavin T (TT); (**A**) and methylene blue (MB); (**B**) sorption by dried biomass of freshwater moss *V. dubyana* ($C_B = 0.5$ g/dm³) from single sorption systems after 2 h interactions at 25 °C and pH 6.0.

The comparison of maximal sorption capacities Q_{max} determined from Langmuir model with those of other authors is reported in Table 3. Although the data about TT (bio)sorption are very limited, our results indicate that *V. dubyana* biomass exhibited the highest Q_{max} in comparison with other organic and inorganic sorbents. Similarly, in case of MB biosorption the moss biomass has a high Q_{max} when compared with different sorbents of plant origin.

Table 3. Comparison of Q_{max} values of different sorbents determined from Langmuir isotherm for TT and MB.

Sorbent	$Q_{max\ MB}$ (mg/g)	$Q_{max\ TT}$ (mg/g)	pH	T (°C)	Reference
V. dubyana	229	119	6.0	25	present study
hop leaf biomass	184	77.6	6.0	25	[27]
Rhytidiadelphus squarrosus	-	98.9	4.0	25	[34]
montmorillonite *	-	95.2	6.0	25	[35]
Fomitopsis carnea	-	21.9	-	30	[14]
Corn husk	47.95	-	6.0	25	[8]
Oak acorn peel	120.5	-	7.0	24	[36]
Salvia miltiorrhiza	100	-	7.0		[12]
Eucalyptus bark	204.8	-	9.9	30	[37]
Platanus leaf	99.1	-	7.0	30	[38]
Banana stalk	243.9	-	-	30	[39]

* Cetylpyridinium modified.

In comparison with single sorption systems, in binary or multicomponent mixtures cationic dyes may interact or compete for binding sites of moss biomass. Therefore, the behavior of each species in a multicomponent system depends strongly on the number and properties of other species present. In addition, the solution pH, the physical and chemical properties of both the sorbent and sorbate significantly influenced the sorption process. The simultaneous Langmuir and Freundlich sorption isotherms of MB and TT from binary system MB + TT (concentration ratio $C_{0\ TT}:C_{0\ MB} = 1:1$) by *V. dubyana* biomass are shown in Figure 6A,B. Table 4 presents isotherm parameters calculated by non-linear regression analysis.

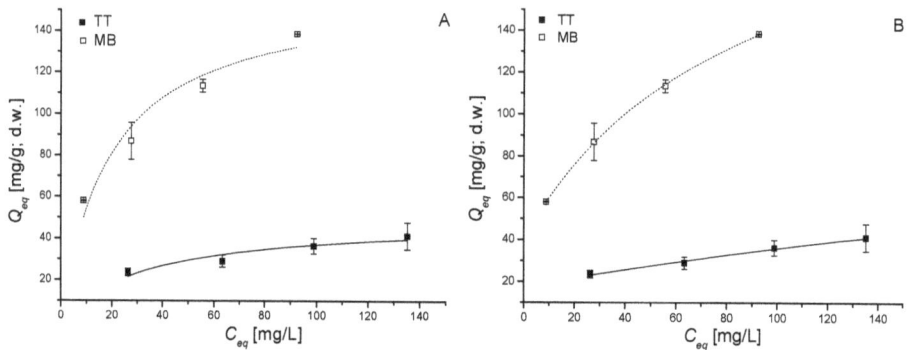

Figure 6. Fit of the Langmuir (**A**) and Freundlich (**B**) isotherms of thioflavin T (TT) and methylene blue (MB) biosorption by dried biomass of freshwater moss *V. dubyana* (C_B = 0.5 g/dm^3) from binary sorption system TT + MB (concentration ratio $C_{0\,TT}$:$C_{0\,MB}$ = 1:1) after 2 h interactions at 25 °C and pH 6.0.

Table 4. Langmuir and Freundlich equilibrium parameters (±SD) obtained by non-linear regression analysis for TT and MB biosorption by *V. dubyana* biomass in binary system TT + MB.

Dye	Langmuir			Freundlich		
	Q_{max} (mg/g)	b (L/mg)	R^2	K (L/g)	$1/n$	R^2
TT [1]	49.3 ± 6.2	0.03 ± 0.01	0.905	6.92 ± 1.46	2.77 ± 0.36	0.972
MB [1]	160 ± 17	0.05 ± 0.02	0.950	25.4 ± 0.72	2.67 ± 0.05	0.999
TT + MB [2]	232 ± 29	0.013 ± 0.004	0.955	16.7 ± 1.9	2.29 ± 0.12	0.839

[1] Sorption of individual dye in binary system TT + MB. [2] Total TT and MB sorption ($Q_{eq} = Q_{eq}$(TT) + Q_{eq} (MB)) in binary system TT + MB.

It is evident that sorption of both TT and MB increased with increasing of dyes concentrations in binary mixture. However, the coexistence of MB and TT influenced the sorption capacity of each other, advocating the competitive sorption between TT and MB. The sorption of MB is less affected by the presence of TT than the sorption of TT by the presence of MB in solution.

Maximum sorption capacities Q_{max} of cationic dyes in binary system TT + MB calculated from Langmuir model reached 49.3 ± 6.2 mg/g (TT) and 160 ± 17 mg/g (MB) and values of affinity parameter b (0.03 ± 0.01 L/mg for TT; 0.05 ± 0.02 L/mg for MB) indicating markedly higher affinity of moss biomass to MB. However, both Q_{max} values in binary system are significantly lower in comparison with Q_{max} values obtained in single sorption systems (Table 1). The sorption isotherms of total dye sorption (data calculated as $Q_{eq} = Q_{eq}$(TT) + Q_{eq}(MB); $C_{eq} = C_{eq}$(TT) + C_{eq}(MB)) were also constructed and evaluated (not shown). From Table 4 it is clear that maximum sorption capacity Q_{max} for MB from single sorption system calculated from Langmuir model (Table 2) is almost equal to total Q_{max}(TT + MB) in binary system. This confirms the hypothesis that both TT and MB in binary mixture TT + MB are sequestered by identical binding sites on biomass surface and observed differences in individual dye sorption capacities Q are due to (i) mutual competitive effects between dyes and (ii) different affinity of biomass to cationic dyes. Remenárová et al. [19] described competitive effects during sorption in binary system TT + MG (malachite green) by moss *R. squarrosus*. They revealed that TT significantly affected biosorption of MG in binary system TT + MG, the competitive effect of MG on TT was less pronounced. On the contrary, Albadarin et al. [18] observed that the total adsorbed quantity of single dyes is only slightly larger than a mixture of two components with the same concentration in binary system MB and alizarin red S (ARS). This indicating the presence of limited number of active sites by which the two dyes can be sequestrated and for which they will, to some extent, compete for in binary system. Based on obtained data and results of other authors we suppose that variance in affinity in multicomponent dye systems could be attributed to the different chemical structure of dye molecules and physicochemical properties of sorbent used. Tabaraki et al. [28] stated that the chemical structure of dye molecules, the number of sulfonic groups, the basicity and molecular weight of molecules significantly influenced biosorption capacity and affinity.

4. Conclusions

In the present study, dried biomass of freshwater moss *V. dubyana* has been used as biosorbent for cationic dyes methylene blue and thioflavin T removal from both single and binary systems. Results revealed that an increase in pH has a positive effect on TT and MB sorption and the electrostatic attraction is one of the possible mechanism of MB and TT removal by moss *V. dubyana*. The experimental equilibrium biosorption data of TT and MB from single systems were well described by Langmuir isotherm and maximum sorption capacities Q_{max} were 119 ± 11 mg/g for TT and 229 ± 9 mg/g for MB. In binary mixture, the presence of MB caused significant decrease of TT sorption indicating higher affinity of biomass to MB. We conclude that in binary systems the competitive effects can substantially influence the sorption process and should be thoroughly evaluated before application of selected adsorbents for removal of basic dyes from wastewaters containing mixtures of dyes.

Acknowledgments: This work was supported by the project of the Cross-border Co-operation Programme and co-financed with European Regional Development Fund (ERDF), the grant number HUSK/1101/1.2.1/0148 as well as the project of the Operational Program Research and Development and co-financed with European Regional Development Fund (ERDF), the grant number ITMS 26220220191.

Author Contributions: Martin Pipíška and Miroslav Horník conceived and designed the experiments; Denisa Partelová and Martin Valica performed the experiments; Denisa Partelová and Miroslav Horník analyzed the experimental data; Juraj Lesný and Stanislav Hostin contributed materials and analysis tools; Martin Pipíška and Miroslav Horník wrote the paper.

Conflicts of Interest: The authors declare no conflict of interest.

References

1. Ghaly, A.E.; Ananthasankhar, R.; Alhattab, M.; Ramakrishnan, V.V. Production, characterization and treatment of textile effluents: A critical review. *J. Chem. Eng. Process Technol.* **2014**, *5*, 182. [CrossRef]

2. Market Research Future. *Global Textile Dyes Market Research Report—Forecast to 2023*; ID: MRFR/CnM/2238-HCRR; Market Research Future: Maharashtra, India, 2017.

3. GosReports. Global Textile Dyes Industry 2015 Market Research Report. 2015. Available online: https://www.prnewswire.com/news-releases/global-textile-dyes-industry-report-2015---forecasts-to-2020-498532981.html (accessed on 2 April 2015).

4. Pereira, L.; Alves, M. Dyes—Environmental impact and remediation. In *Environmental Protection Strategies for Sustainable Development*; Malik, A., Grohmann, E., Eds.; Springer Science + Business Media B.V.: Dordrecht, The Netherlands, 2012; pp. 111–162, ISBN 978-94-007-1591-2.

5. Dias, A.A.; Sampaio, A.; Bezerra, R.M. Environmental Applications of Fungal and Plant Systems: Decolourisation of Textile Wastewater and Related Dyestuffs. In *Environmental Bioremediation Technologies*; Singh, S.N., Tripathi, R.D., Eds.; Springer: Berlin/Heidelberg, Germany, 2007; pp. 445–463, ISBN 978-3-540-34790-3.

6. Sing, K.; Arora, S. Removal of synthetic textile dyes from wastewaters: A critical review on present treatment technologies. *Crit. Rev. Environ. Sci. Technol.* **2011**, *41*, 807–878. [CrossRef]

7. Robinson, T.; McMullan, G.; Marchant, R.; Nigam, P. Remediation of dyes in textile effluent: A critical review on current treatment technologies with a proposed alternative. *Bioresour. Technol.* **2001**, *77*, 247–255. [CrossRef]

8. Paşka, O.M.; Pəcurariu, C.; Muntean, S.G. Kinetic and thermodynamic studies on methylene blue biosorption using corn-husk. *RSC Adv.* **2014**, *4*, 62621–62630. [CrossRef]

9. Daneshvar, E.; Vazirzadeh, A.; Niazi, A.; Sillanpää, M.; Bhatnagar, A. A comparative study of methylene blue biosorption using different modified brown, red and green macroalgae—Effect of pretreatment. *Chem. Eng. J.* **2017**, *307*, 435–446. [CrossRef]

10. Liang, J.; Xia, J.; Long, J. Biosorption of methylene blue by nonliving biomass of the brown macroalga *Sargassum hemiphyllum*. *Water Sci. Technol.* **2017**, *76*, 1574–1583. [CrossRef] [PubMed]

11. Horník, M.; Šuňovská, A.; Partelová, D.; Pipíška, M.; Augustín, J. Continuous sorption of synthetic dyes on dried biomass of microalga *Chlorella pyrenoidosa*. *Chem. Pap.* **2013**, *67*, 254–264. [CrossRef]

12. Zhao, S.; Zhou, T. Biosorption of methylene blue from wastewater by an extraction residue of *Salvia miltiorrhiza* Bge. *Bioresour. Technol.* **2017**, *219*, 330–337. [CrossRef] [PubMed]

13. Liu, J.; Li, E.; You, X.; Hu, C.; Hu, Q. Adsorption of methylene blue on an agro-waste oiltea shell with and without fungal treatment. *Sci. Rep.* **2016**, *6*, 38450. [CrossRef] [PubMed]

14. Maurya, N.S.; Mittal, A.K. Biosorptive uptake of cationic dyes from aqueous phase using immobilised dead macro fungal biomass. *Int. J. Environ. Technol. Manag.* **2011**, *14*, 282–293. [CrossRef]

15. Rizzi, V.; D'Agostino, F.; Fini, P.; Semeraro, P.; Cosma, P. An interesting environmental friendly cleanup: The excellent potential of olive pomace for disperse blue adsorption/desorption from wastewater. *Dyes Pigment.* **2017**, *140*, 480–490. [CrossRef]

16. Rizzi, V.; D'Agostino, F.; Gubitosa, J.; Fini, P.; Petrella, A.; Agostiano, A.; Semeraro, P.; Cosma, P. An Alternative use of olive pomace as a wide-ranging bioremediation strategy to adsorb and recover disperse orange and disperse red industrial dyes from wastewater. *Separations* **2017**, *4*, 29. [CrossRef]

17. Semeraro, P.; Rizzi, V.; Fini, P.; Matera, S.; Cosma, P.; Franco, E.; García, R.; Ferrándiz, M.; Núñez, E.; Gabaldón, J.A.; et al. Interaction between industrial textile dyes and cyclodextrins. *Dyes Pigment.* **2015**, *119*, 84–94. [CrossRef]

18. Albadarin, A.B.; Mangwandi, C. Mechanism of Alizarin red S and Methylene blue biosorption onto a olive stone by-product: Isotherm study in single and binary systems. *J. Environ. Manag.* **2015**, *164*, 86–93. [CrossRef] [PubMed]

19. Remenárová, L.; Pipíška, M.; Horník, M.; Augustín, J. Biosorption of cationic dyes BY1, BY2 and BG4 by moss *Rhytidiadelphus squarrosus* from binary solutions. *Nova Biotechnol.* **2009**, *9*, 239–247.

20. Giwa, A.R.A.; Abdulsalam, K.A.; Wewers, F.; Oladipo, M.A. Biosorption of acid dye in single and multidye systems onto sawdust of locust bean (*Parkia biglobosa*) tree. *J. Chem.* **2016**, *2016*, 6436039. [CrossRef]

21. Hoagland, D.R. Optimum nutrient solution for plants. *Science* **1920**, *52*, 562–564. [CrossRef] [PubMed]

22. Zhang, Y.; Liu, W.; Xu, M.; Zheng, F.; Zhao, M. Study of the mechanisms of Cu^{2+} biosorption by ethanol/caustic-pretreated baker's yeast biomass. *J. Hazard. Mater.* **2010**, *178*, 1085–1093. [CrossRef] [PubMed]

23. Ncibi, M.C.; Ben Hamissa, A.M.; Fathallah, A.; Kortas, M.H.; Baklouti, T.; Mahjoub, B.; Seffen, M. Biosorptive uptake of methylene blue using Mediterranean green alga *Enteromorpha* spp. *J. Hazard. Mater.* **2009**, *170*, 1050–1055. [CrossRef] [PubMed]

24. Hameed, B.H. Grass waste: A novel sorbent for the removal of basic dye from aqueous solution. *J. Hazard. Mater.* **2009**, *166*, 233–238. [CrossRef] [PubMed]

25. Fernandez, M.E.; Nunell, G.V.; Bonelli, P.R.; Cukierman, A.L. Batch and dynamic biosorption of basic dyes from binary solutions by alkaline-treated cypress cone chips. *Bioresour. Technol.* **2012**, *106*, 55–62. [CrossRef] [PubMed]

26. Salazar-Rabago, J.J.; Leyva-Ramos, R.; Rivera-Utrilla, J.; Ocampo-Perez, R.; Cerino-Cordova, F.J. Biosorption mechanism of Methylene Blue from aqueous solution onto White Pine (*Pinus durangensis*) sawdust: Effect of operating conditions. *Sustain. Environ. Res.* **2017**, *27*, 32–40. [CrossRef]

27. Partelová, D.; Šuňovská, A.; Marešová, J.; Horník, M.; Pipíška, S.; Hostin, S. Removal of contaminats from aqueous solutions using hop (*Humulus lupulus* L.) agricultural by-products. *Nova Biotechnol. Chim.* **2015**, *14*, 212–227. [CrossRef]

28. Tabaraki, R.; Sadeghinejad, N. Biosorption of six basic and acidic dyes on brown alga *Sargassum ilicifolium*: Optimization, kinetics and isotherm studies. *Water Sci. Technol.* **2017**, *75*, 2631–2638. [CrossRef] [PubMed]

29. Tural, B.; Ertaş, E.; Enez, B.; Fincan, S.A.; Tural, S. Preparation and characterization of a novel magnetic biosorbent functionalized with biomass of *Bacillus subtilis*: Kinetic and isotherm studies of biosorption processes in the removal of Methylene Blue. *J. Environ. Chem. Eng.* **2017**, *5*, 4795–4802. [CrossRef]

30. Kumar, K.V.; Porkodi, K. Mass transfer, kinetics and equilibrium studies for the biosorption of methylene blue using *Paspalum notatum*. *J. Hazard. Mater.* **2007**, *146*, 214–226. [CrossRef] [PubMed]

31. Uddin, N.; Islam, T.; Das, S. A novel biosorbent, water-hyacinth, uptaking methylene blue from aqueous solution: Kinetics and equilibrium studies. *Int. J. Chem. Eng.* **2014**, *2014*, 819536. [CrossRef]

32. Zhu, L.; Wang, Y.; Zhu, F.; You, L.; Shen, X. Evaluation of the biosorption characteristics of *Tremella fuciformis* for the decolorization of cationic dye from aqueous solution. *J. Polym. Environ.* **2017**, in press. [CrossRef]

33. Hamitouche, A.; Haffas, M.; Boudjemaa, A.; Benammar, S.; Sehailia, M.; Bachari, K. Efficient biosorption of methylene blue, malachite green and methyl violet organic pollutants on biomass derived from *Anethum graveolens*: An eco-benign approach for wastewater treatment. *Desalin. Water Treat.* **2017**, *5*, 225–236. [CrossRef]

34. Remenárová, L.; Pipíška, M.; Horník, M.; Augustín, J. Sorption of cationic dyes from aqueous solution by moss *Rhytidiadelphus squarrosus*: Kinetics and nequilibrium study. *Nova Biotechnol.* **2009**, *9*, 75–84.

35. Shin, W.S. Competitive sorption of anionic and cationic dyes onto cetylpyridinium-modified montmorillonite. *J. Environ. Sci. Health A Toxic Hazard. Subst. Environ. Eng.* **2008**, *43*, 1459–1470. [CrossRef] [PubMed]

36. Kuppusamy, S.; Venkateswarlu, K.; Thavamani, P.; Lee, Y.B.; Naidu, R.; Megharaj, M. *Quercus robur* acorn peel as a novel coagulating adsorbent for cationic dye removal from aquatic ecosystems. *Ecol. Eng.* **2017**, *101*, 3–8. [CrossRef]

37. Afroze, S.; Sen, T.K.; Ang, M.; Nishioka, H. Adsorption of methylene blue dye from aqueous solution by novel biomass *Eucalyptus sheathiana* bark: Equilibrium, kinetics, thermodynamics and mechanism. *Desalin. Water Treat.* **2016**, *57*, 5858–5878. [CrossRef]

38. Kong, L.; Gong, L.; Wang, J. Removal of methylene blue from wastewater using fallen leaves as an adsorbent. *Desalin. Water Treat.* **2015**, *53*, 2489–2500. [CrossRef]

39. Hameed, B.H.; Mahmoud, D.K.; Ahmad, A.L. Sorption equilibrium and kinetics of basic dye from aqueous solution using banana stalk waste. *J. Hazard. Mater.* **2008**, *158*, 499–506. [CrossRef] [PubMed]

environments

MDPI

Article

Biological Treatment by Active Sludge with High Biomass Concentration at Laboratory Scale for Mixed Inflow of Sunflower Oil and Saccharose

Pedro Cisterna

Department of Civil and Environmental Engineering, University of Bío Bío, Concepción, Z.C. 378000, Chile; pcisterna@ubiobio.cl; Tel.: +56-949-807-764

Received: 17 July 2017; Accepted: 25 September 2017; Published: 28 September 2017

Abstract: We studied and quantified the elimination of sunflower oil from a wastewater influent using a biological treatment by activated sludge. Estimation of the biodegraded material was obtained doing a mass balance, and we conducted a follow-up of the different operational parameters and design. We delivered information about the operation of a system for treatment by activated sludge fed with an influent with sunflower oil and saccharose. The influent was previously agitated before entering the effluent sludge in a lab-scale plant. The working range for oil concentration was 100 to 850 mg/L in the influent. Biodegradation was in the range of 60% to 51%. The process works better with a high initial concentration of biomass (7500 mg/L) in order to absorb the impacts caused by the oil on the microorganisms. The lowest total suspended solids concentration was 4500 mg/L. The elimination of sunflower oil in biodegradation and flotation was on the order of 90%.

Keywords: biodegradation; fat; activated sludge

1. Introduction

Domestic wastewater and some types of industrial wastewater contain fats and oils in a considerable proportion. The fraction of lipids in urban wastewater is 30–40% of the chemical oxygen demand (COD), a test that measures organic matter [1].

The typical solution used for the elimination of fats and oils is based on physical and physico-chemical treatments, generating high volumes of fatty semisolid waste that is also highly undesirable from an environmental point of view. Considering the above mentioned, it is worth the effort to investigate an alternative biologically-based treatment for such types of waste.

In this laboratory-scale experiment, artificial wastewater containing oil and a substrate simulating residual water of urban or industrial origin, is biologically treated by activated sludge.

Degradation of sunflower oil was studied in an influent that also had saccharose. The influent is treated in an activated sludge plant designed to work with high concentrations of biomass. Initial biomass concentration (TSS) was 7500 mg/L, in order to face the situation of a substrate of a slower biodegradation.

The behavior of lipids in active sludge processes is, in general, not well understood. The literature generally states that lipids and fatty acids can be removed by biological treatment, which eventually causes foam formation composed of filamentous bacteria and flocs that also inhibit microbial reproduction [2].

In general, flotation is used in the separation of solids from fluids or between immiscible fluids. There are three different types of flotation, the difference is based on how the air is introduced into the wastewater: by the use of air at atmospheric pressure or dissolved and induced via air [3].

Physical-chemical treatment is applied to wastewater with fats and oils highly emulsified and dispersed, and where the size of fat particles is less than 20 microns. When using a dissolved air flotation system to treat wastewater in the oil industry, the elimination efficiency is increased from 50% to 88% due to the addition of chemicals [4]. On the other hand, using electro-coagulation achieves high effectiveness through the destabilization of the emulsions, improving the removal of oils and fats [5].

As for the biodegradability of fats and oils, there is generic information classifying them as a slowly biodegradable substance. As for the degradation itself, bacteria initially saves these substances in their cytoplasm and later, through an enzymatic process, they perform hydrolysis to produce an assimilable substrate that can be biodegraded [6].

There are three types of reactions catalyzed by microbial enzymes: oxidatives, hydrolytic, and synthesis. The hydrolytic enzymes are used to hydrolyze insoluble complex compounds, such as fats and oils, on simple components that pass through the cellular membrane by diffusion. These enzymes (for instance oxidoreductases) act outside the cell wall [7].

Fats and oils are biodegraded by a wide range of microorganisms, however, the most common is an extracellular enzyme called lipases. This enzyme releases fatty acids, as a result of enzymatic action [8]. Fatty acids can be biodegraded by a wider range of microorganisms, including microorganisms that do not produce extracellular lipolytic enzymes [9].

Wastewater with high lipid concentration inhibits the activity of microorganisms in biological treatment systems, such as activated sludge and methane fermentation. To reduce such inhibitory effects, microorganisms capable of effectively degrading edible oils can be selected from different environmental sources [10].

Taking into account that the fat and oil biodegradation process is slow, they do not enhance the development and growth of bacterial colonies and, therefore, it is recommended to acclimatize the biomass to this type of substrate for a certain period of time, such that the biomass adapts and achieves the expected biodegradation efficiency. The adaptation to the new substrate is due to mutations or changes caused by environmental conditions and chemical or physical agents, which modify the cell DNA, giving it new characteristics, which allows the cell to degrade new substrates from those generated by biodegradation. Spontaneous mutations occur in one of 10^6 cells; however, the molecule DNA is able to self-replicate [11].

The process of activated sludge has been used for wastewater treatment from industrial and urban sources for a century. The design of these plants is carried out mostly based on empirical evidence. Since the 1960s, a more rational solution for the design of activated sludge has been developed. The latter solution is based on the observation that if any urban or industrial wastewater is put through an aeration process for a sufficient period of time, the content of organic matter is reduced, creating, at the same time, a flocculent sludge [12].

The typical design parameter of the activated sludge treatment system is the mass load (ML). By definition, the mass load of the aeration tank is the relation between the daily feed mass of organic matter that gets into the aeration tank and the degrading biomass content inside the same tank [7].

There are many studies about the behavior of lipids in biological treatment assessing the disposal process [5]. Biodegradation of substances, slightly or not soluble in water, is one of the greatest problems in the use of biotechnology for the treatment of contaminated solid and liquid waste [13]. Microorganisms use a wide variety of organic compounds, like carbon as an energy source, for growth. When these substrates are not reachable due to low solubility, competition with other microorganisms, or other environmental factors, the use of biosurfactants is recommended to obtain a carbon source [14].

In a biological treatment where activated sludge is in use, the efficient contact of phases, water and oil, is very important and, therefore, a significant interfacial area is required to reduce mass transfer limitations. The interfacial area can be expanded by delivering energy to the system by means of mechanical stirring or an electric field. The increment of the interfacial surface between the aqueous phase and oily or fatty phases is often implemented by mechanical stirring [15].

Regarding the stirred aeration tank used to complete the mixing and the corresponding dispersion level, there are two important factors that largely determine the emulsion level: bubble size and distribution, and the fraction of the dispersed phase. The average bubble size is between 150 μm and 250 μm [16], which can be obtained by means of a suitable booster and fine bubble diffusers.

If there is a suitable enzyme concentration and an optimal interfacial area between the aqueous phase and the oily phase, the mass transfer is solved and it gives way to the hydrolysis stage [17].

This investigation studies and quantifies sunflower oil biodegradation using a biological treatment system based on activated sludge at the laboratory scale with a high initial concentration of biomass (7500 mg/L).

2. Materials and Methods

2.1. The Choice of Sunflower Oil Is Explained Based in the Following Criteria

- It is the most frequent oil used in Chile;
- It is well standardized; and
- It is a highly accessible product.

The chemical composition of sunflower oil is shown in Table 1.

Table 1. Fatty acids that make up sunflower oil and specific stereoanalysis [18]. Results in % moles.

Fat	Position	16:0	18	18:1(9)	18:2(9,12)	18:3(9,12,15)
	1	10.6	3.3	16.6	69.5	-
Sunflower oil girasol	2	1.3	1.1	21.5	76.0	-
	3	9.7	9.2	27.6	53.5	-

2.2. Physico-Chemical Parameters and Analytical Methods

- Chemical Oxygen Demand (COD)

The potassium dichromate method was used to evaluate COD levels. The method used is a variation of the standard method [19], however, it maintains the basis of it. The variation used has the advantage that it uses a significantly smaller sample and reagents. The sample is chemically oxidized through the action of potassium dichromate at 150 °C for two hours. Silver sulfate is used as a catalyst and mercury sulfate is used to avoid possible interferences with chloride. Afterwards, determination by spectrophotometry at 600 nm is performed. Equipment and instruments used to determine the various parameters to characterize the wastewater were used.

The following parameters with their corresponding methods were measured:

- Total Suspended Solids (TSS), Volatile Suspended Solids (VSS) and Sludge Volumetric Index (SVI)

TSS is determined by filtering a known volume of the sample on Whatman (Whatman plc, Maidstone, UK) 4.7 cm GF/C glass fiber filters and then drying it at 103–105 °C. The difference in weight of the filter before and after filtration is used to estimate the TSS, 209C method [20]. The Volatile Suspended Solids are determined by weight loss after calcination at 550 °C, 208E method [20]. For Sludge Volumetric Index, method 213E is used [20].

- Fats and Oils

When determining the fats and oils, the Gravimetric Assay Soxhlet method is used. This method quantifies substances with similar characteristics on the basis of their common solubility in an appropriate solvent, 213E method [20].

2.3. Determination of Chemical Oxigen Demand (COD)-Substrate Relationships

Samples composed by mixtures of water and substrates prepared at different concentrations, and their respective COD is estimated. This test is performed in order to produce a calibration curve and establish a ratio (substrate concentration/COD).

2.4. Continuous Equipment

In the current investigation, an activated sludge plant at the laboratory scale is used to conduct biodegradability tests in wastewater with oils and fats. To meet these objectives, experimental work is required, with such parameter information describing the process dynamics regarding the aeration and sedimentation tanks regarding fat and oil content of the wastewater. For this purpose, BIOCONTROL-MARK 2 equipment was used. The details of the equipment is shown in Figure 1.

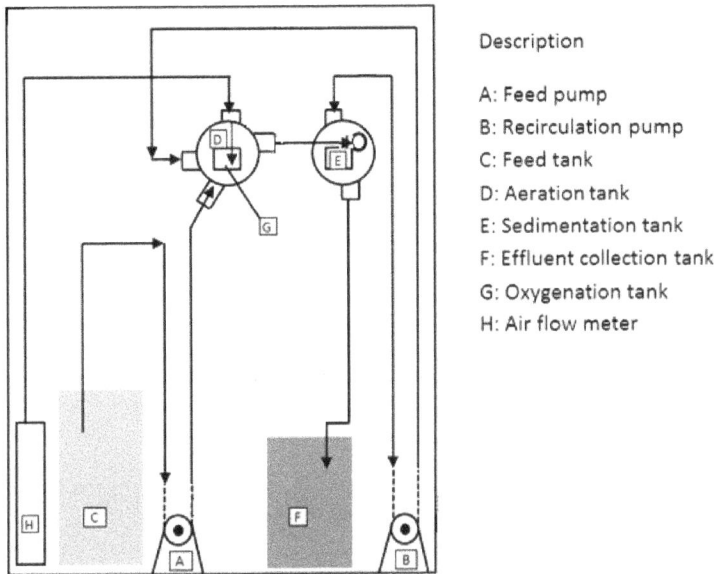

Description

A: Feed pump
B: Recirculation pump
C: Feed tank
D: Aeration tank
E: Sedimentation tank
F: Effluent collection tank
G: Oxygenation tank
H: Air flow meter

Figure 1. Experimental equipment diagram.

This experimental equipment consists essentially of the following parts:

● Control Unit

Composed of a main switch, an air cylinder, and is complemented with a flow meter and a flow regulation system. Additionally, a wastewater feed pump, complemented with a flow rate regulation system, a timer for intermittent operations, and an ON-OFF switch allowing sludge recycling from the sedimentation tank to the aeration tank.

● Aeration Tank

It consists of a transparent Plexiglas® (Vittadini Riferimenti, Milan, Italy) cylinder with a height of 38 cm and a diameter of 20 cm, which has outlets at various heights associated with different volumes (7, 8, 9 and 10 L). There are two separated inlets allowing the recirculation of sludge from the top. The influent to be treated is placed at the bottom. In addition, the system has two ceramic diffusers placed in the bottom in a way that they can disperse the air in tiny bubbles [21].

● Sedimentation Tank

This consists of a transparent Plexiglas® (Vittadini Riferimenti, Milan, Italy) cylinder, where its lower part is cone-shaped to make sludge sedimentation and thickening easier.

The mixed liquor is fed from the aeration tank, which has, in its upper part, an outlet. This flow escapes by overflow, when it arrives to the sedimentation tank. The solid phase decantation gives a method to a downward flow. The decanted sludge is separated and recirculated at the bottom through a pump towards the aeration tank. Treated water also uses the overflow mechanism to be evacuated to the storage tank.

2.5. Experimental Methodology

a.   Feed Preparation

The treatment system was fed initially with synthetic wastewater prepared in the laboratory according to the typical characteristics of strong urban wastewater [22]. This wastewater has an approximate BOD of 400 mg/L, with the corresponding proportions of nitrogen and phosphorus, in a relation of BOD:N:P = 100:5:1. Approximately 400 mg of saccharose, 20 mg of phosphate hydrogen of potassium, and 100 mg of ammonium chloride were added per liter of water. Measurements begin when sunflower oil is added, the concentration of this substrate gradually increases. Feed was prepared daily and nitrogen and phosphorus increased according to the organic input coming from fats and oils.

b.   Operating Modes

The synthetic wastewater was poured into a storage pond of approximately 50 L, in which a stirring unit has been installed to disperse the oil or fat. Through a peristaltic pump, controlled by the control unit, it drives the feed to the aeration tank. Oxygen feed and recirculation flow are controlled by the control unit. Process effluent is collected in a 30 L volume tank, where the samples are taken to be processed. The flow of synthetic wastewater is 25 L/day.

3. Results

3.1. COD-Substrate Relationships

From the experimental values, a straight line with a slope of 2.13 is obtained, as shown in Figure 2, from which it can be stated that this type of oil has a COD per gram, which is above of other organic substances [23]. The model obtained is: Y = 2.1255X.

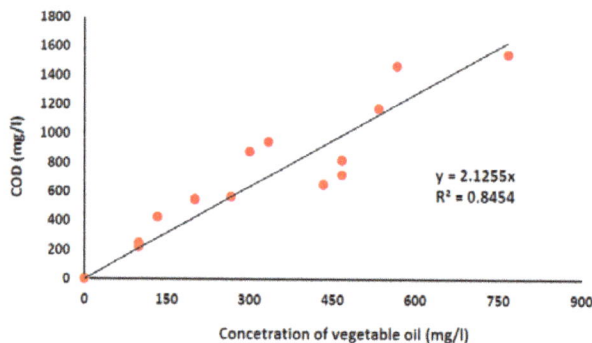

Figure 2. COD-sunflower oil relationship.

From the experimental values, a straight line with a slope of 1.17 is obtained, as shown in Figure 3, from which it can be stated that the saccharose has a COD per gram, which is above of other organic substances [23]. The model obtained is: Y = 1.1744X.

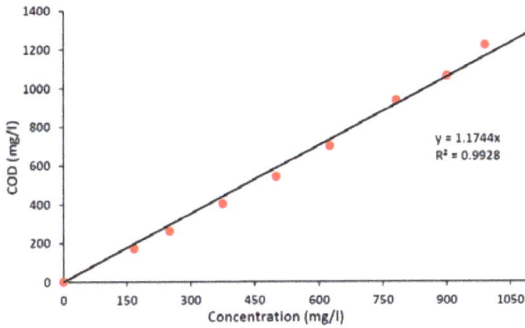

Figure 3. COD-saccharose relationship.

3.2. Mass Balance

In this experience, water and oil are mixed by mechanical stirring. Even though part of the added oil remains accumulated, it is important to determine the oil fraction which does not enter the aeration tank.

Biodegradability is estimated considering the oil that actually enters the aeration tank, which will correspond to the oil that is subject to a biological treatment process using activated sludge.

The influent has saccharose and sunflower oil. Saccharose concentration is constant and sunflower oil concentration is gradually increased. In this case, the initial biomass concentration present in the aerobic reactor is 7500 mg/L.

As shown in Figure 4: F0, C0: Flow and concentration of oil entering the mixing tank. F1, C1: Flow and concentration of oil from the feed tank to the aeration tank. F2, C2: Flow and concentration of fat and oil exiting the aeration tank. F3, C3: Flow and concentration of fat and oil from the purified effluent. F4, C4: Flow and concentration of fat and oil in the recirculation flow. M1: Mass of oil contained in the mixing liquor of the aeration tank. M2: Mass of oil floating on top of the sedimentation tank and the mass of oil and fat at the bottom of sedimentation tank by adherence to the biomass.

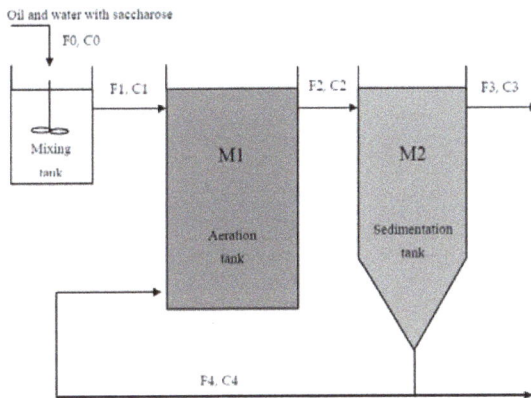

Figure 4. The plant process of activated sludge diagram.

Matter Balance: Determination of Biodegradability

From the matter balance, the mass of biodegraded sunflower oil is obtained, which is part of the influent to the system, as follows:

For the aeration tank:

$$dM_1/dt = F_1C_1 - F_2C_2 - r_AV + F_4C_4 \tag{1}$$

For the secondary sedimentation tank:

$$dM_2/dt = F_2C_2 - F_3C_3 - F_4C_4 - F_6C_6 \tag{2}$$

It is pertinent to point out that, $F_6 = 0$.

M_1: mass of oil and fat in the aeration tank.

M_2: mass of oil and fat in the sedimentation tank.

V: Volume of reactor.

r_A: Disappearance speed of vegetable oil.

The balance for the whole system is:

$$dM_1/dt + dM_2/dt = F_1C_1 - F_3C_3 - r_AV \tag{3}$$

From this expression, we have that the term r_AV, corresponding to vegetable oil disappearing by unit of time, is ultimately the oil purely degraded by the microorganisms, such that:

$$r_AV = F_1C_1 - F_3C_3 - dM_1/dt - dM_2/dt \tag{4}$$

As it has been said, a significant part of the oil floats and, therefore, it is not biodegraded and goes directly into the secondary sedimentation tank, where it is accumulated. By reasons of technical feasibility, the enforcement of the matter balance must be performed holistically and not derivatively. Therefore, this accumulation effect due to floating is considered and measured, which is very important in order to measure the level of biodegradability corresponding to the influent that has oil.

When the matter balance is performed holistically, for a determined period of time, we have:

$$M_1 + M_2 = F_1C_1Dt - F_3C_3Dt - r_ADt \tag{5}$$

where Dt is the period of time, (one day).

Then, if a subscript is used to characterize the corresponding day, we have:

For day 1:

$$M_{11} + M_{21} = F_{11}C_{11}Dt - F_{31}C_{31}Dt\text{-}Vr_ADt \tag{6}$$

For day 2:

$$M_{12} + M_{22} = F_{12}C_{12}dt - F_{32}C_{32}Dt - r_AVDt \tag{7}$$

For day n:

$$M_{1n} + M_{2n} = F_{1n}C_{1n}Dt - F_{3n}C_{3n}Dt - r_AVDt \tag{8}$$

the mass balance for a determined amount of days operating:

$$\Sigma(M_{1i} + M_{2i}) = F_{1i}Dt\Sigma(C_{1i}) - F_{3i}Dt\Sigma(C_{3i}) - r_AVDt \tag{9}$$

Therefore, the expected biodegraded oil can be estimated by:

$$F_{1i}Dt\Sigma(C_{1i}) - F_{3i}Dt\Sigma(C_{3i}) - \Sigma(M_{1i} + M_{2i}) = r_AVDt \tag{10}$$

Then biodegradability is:

$$D = r_AVDt/F_{1i}Dt\Sigma(C_{1i})$$

$$D = \{F_{1i}Dt\Sigma(C_{1i}) - F_{3i}Dt\Sigma(C_{3i}) - \Sigma(M_{1i} + M_{2i})\}/F_1Dt\Sigma(C_{1i}) \tag{11}$$

3.3. Experimental Protocol of Matter Balance of the Activated Sludge Plant

In this protocol an experimental method to conduct a thorough and accurate balance of fat and oil matter is proposed. From the process diagram, the streams of input and output are defined and measured both in flow and concentration.

Feed flow and concentration of fat and oil (F_1, C_1) are known and set according to the experimental conditions.

F_2, C_2: It is not significant for the matter balance.

F_3, C_3: This flow is equal to the input flow and, therefore, it is given by the predetermined conditions for the experiment, and C3 concentration is obtained by Soxhlet method.

F_4, C_4: It is not significant for the matter balance.

M_1: Oil concentration in mixed liquor measured by Soxhlet using a sample of 250 mL. The concentration is calculated for the total volume of mixed liquor.

M_2: In this case, sunflower oil mass accumulated in the sedimentation tank in a determined period of time must be estimated, for which it is necessary to remove, dry, and weigh the supernatant oil.

This task was performed by implementing a drainage in the secondary sedimentation tank discharging to an auxiliary tank, and collecting the corresponding part of oil in another container. After this, water evaporation is performed in a water bath and the separated oil is weighed to determine the oil mass accumulated in a given period of time.

M_{21}: This is determined by Soxhlet extraction of oil accumulated in the biomass of the sedimentation tank.

3.4. Matter Balance Results

The empirical results of the mass balance is shown in Table 2.

Table 2. Matter balance of activated sludge with stirring and the increase of the initial biomass.

Period of Operation of the Mass Balance (day)	Fed Oil Mass (g)	Retained Oil Mass (g)	Accumulated Oil Mass (g)	Oil Mass in Effluent (g)	Biodegraded Oil Mass (g)	Biodegradation Efficiency (%)
1–8	24	2	4	4.9	13.1	60
9–15	48	5	13.5	6	23.5	54.6
16–22	94	9	38	5.5	43.5	51

3.5. System Behavior and Operating Parameters

3.5.1. COD Removal and Mass Loading

Figure 5 shows the elimination of COD and its relationship with the mass loading. As a design parameter of the active sludge, it is verified that the biodegradation of the oils and fats decreases with the increase of the mass load. This increase is given by the higher load of fats and oils, which is the substrate of slower biodegradation, and the value of this parameter indicates the accumulation of fats and oils. On the other hand these fats and oils are accumulated in the ponds of the system, therefore, they do not affect the efficiency of the elimination of fats and oils since they do not leave by the effluent.

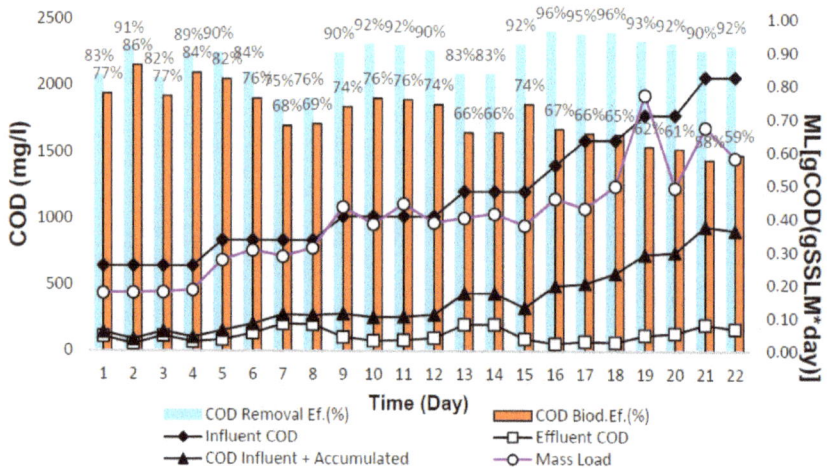

Figure 5. Graphic of the evolution of the removal and biodegradation of COD in the mass load.

3.5.2. Behavior of Biomass

Figure 6 shows the behavior of biomass through total suspended solids. The trend shows a decrease in the concentration of solids as oil is added to the system, although the process is reversed from the ninth day, which is explained by the acclimation of the microorganism to the oil. The high concentration of the initial biomass, 7500 mg/L, allows a resistance to the decline in biomass caused by the fed influent. In fact, the lowest concentration of total suspended solids also corresponds to a high concentration, 4500 mg/L. It is observed that the presence in the influent substrates causes an increase of the SVI of a value of 60 mL/g, to values that exceed 100 mL/g, then are stabilized at an average of 85 mL/g. It is remarkable that, despite the increase experienced by the SVI, the achieved values are within a suitable range, indicating proper biomass sedimentation in the activated sludge.

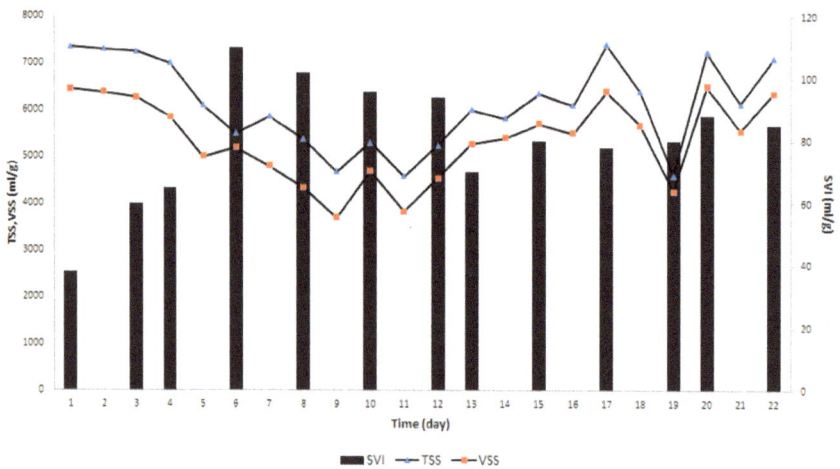

Figure 6. Graphic of the evolution of mass and SVL.

3.5.3. Sunflower Oil Removal and Biodegradation Efficiency

It can be noticed that the concentration of grease and oil in the effluent is independent from the concentration of the influent containing sunflower oil. The latter is because the oil is not biodegraded and it is accumulated in the aeration tank and in the secondary sedimentation tank, as is shown in Figure 7. Therefore, oil is removed by flotation. The elimination of sunflower oil by biodegradation and flotation reaches 90%.

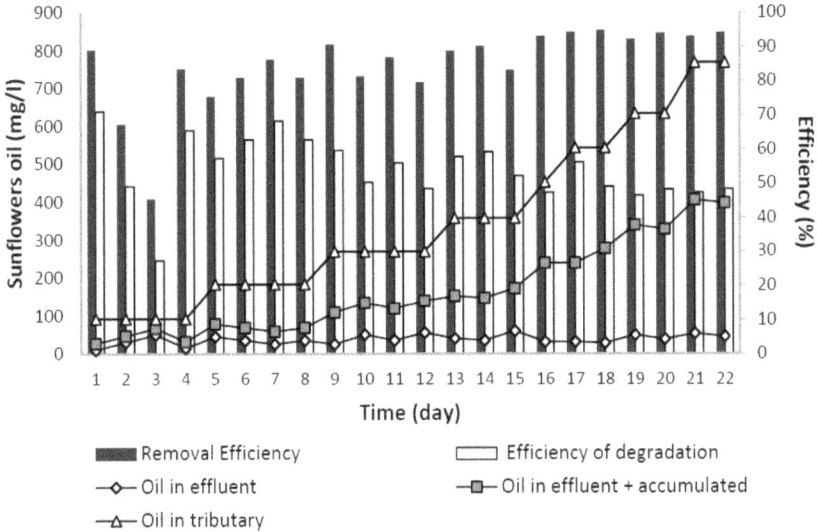

Figure 7. Graphic of the evolution of oil in the influent and effluent, mixed feed, biomass increase, removal efficiency, and biodegradation.

The low solubility of the sunflower oil in the aqueous phase also affects biodegradation, which is mitigated by the emulsification in small drops that create a kind of pseudo-solution corresponding to oil that is feasibly treated biologically.

Another biodegradation limiting factor is the amount of biomass available. In this experience, the oil mass was progressively increased and, therefore, the ratio of biomass oil and organic matter is decreasing. This is reflected by the increase of the mass load, as shown in Figure 5.

In experiments performed to characterize the transformation of lipids in activated sludge under aerobic conditions, the results showed that the total lipid content in the effluent could not be reduced under 300 mg/L when initial content of 2000 mg/L is used [5]. In this experience, despite the presence of an easily biodegradable substrate, such as saccharose, biodegradation reached in excess of 51%, as shown in Figure 7.

4. Discussion

The biodegradation of oils and greases incorporates carbon to the ecological chains through the carbon cycle, which is the most sustainable option.

From the results, a significant level of biodegradation of sunflower oil exceeding 53% is verified. When analyzing this value, the competence of substrates must be considered, since in this case microorganisms have both saccharose and oil as carbon sources, and considering that saccharose is a substrate of easier biodegradation; they are developed with higher biodegradation rates than those species that are compatible with this kind of substrate.

The oil biodegradability level is approximately 53%, on average. This result confirms that, in a significant proportion, oil and fat can be removed from an influent by biodegradation mechanisms of the activated sludge process.

In experiences carried out in a batch reactor with a saccharose and sunflower oil mix, the reached values of biodegradation are on the order of 63–67% for the biomass that, before the experiment, was not subjected to the mix with sunflower oil [24] and, in this case, a continued system of activated sludge, the values go from 51% to 60%, as they have some similarities.

The experience of an activated sludge system working with a biomass concentration within the normal rank (3000–4000 mg/L) shows a significant development of bulking from 120 mL/g to 300 mL/g, approximately [25], and this biomass-reinforced system decreases bulking obtaining an SVI that is between 40 mL/g to 115 mL/g, which shows a drastic difference.

On the other hand, laboratory studies for commercial supplements, multi-species bioaugmentation found removal values of fat and oil from 37% to 62% [26]. This result has some similarities with the effect of sludge acclimatization, which is also of a biological nature.

In the results there is an important percentage of sunflower oil that is not biodegraded; this is due to the lack of solubility of the substrate which originates as a biphasic system which is also increased by the oil tendency to float on water, especially when there is an air flow from the bottom. This means that an important part of the oil does not contact the biomass, violating an essential and basic condition for substrate biodegradation and, therefore, it goes from the feed tank to the secondary sedimentation tank where it is accumulated.

Regarding to the organic matter removal (saccharose and vegetable oil) (Figure 5), if the quality of the effluent resulting from the treatment of activated sludge is considered, optimal levels of performance are achieved, because there is a removal level of COD that is above 90%, which would correspond to the COD being biodegraded, and also to removal by physical processes, such as oil floatation. For aerobic biological treatment, there are experiences in plants at full scale. An evaluation of 55 treatment plants of municipal water in the United States verified that fat and oil concentration in the influent had an average value under 80 mg/L and a BOD of 300 mg/L, and the effluent obtained had a fat and oil concentration under 10 mg/L and a BOD under 40 mg/L to the level of effluent. It also can be observed that the fat and oil removal does not show a seasonal variation [27].

From Figure 5 it can be observed the efficiency of COD removal is over 80% and it does not depend on the mass load applied, which is explained by the fact that part of the removed oil is by physical mechanisms of separation. On the other hand, biodegradation decreases substantially when mass load increases. There are adequate removal levels when it is worked on a range of a conventional regime, but not for high rate ranges.

Therefore, the corresponding value to the COD disposal is higher than the one corresponding to the COD biodegradation because, in the first one, the accumulated oil in the aeration and sedimentation tank is not considered, mainly, which is the COD which is not biodegraded. This is necessary to perform a matter balance of the system so as to allow the more accurate measurement of the organic matter that has actually been biodegraded by the activated sludge.

In the application of biological treatment by activated sludge to the derived effluents from the olive oil industry, the treatment of green wastewater of olive oil-activated sludge was used and the results obtained show that this alternative of treatment is efficient, reaching levels of COD disposal of 75–85%, with a COD of input between 1000 to 1500 mg/L, due mainly to ethanol disposal and fatty acids.

The increase of hydraulic retention time and the temperature improved the sludge removal. Thus, the effluent systematically reached a concentration of 200 to 300 mg/L [28].

These results are very similar to those obtained by other researchers for biodegradation of olive oil in a batch-activated sludge system using both acclimated and non-acclimated sludge. For non-acclimated sludge, levels of biodegradability obtained between day 1 and day 4 were 6% to 68%, while for acclimated sludge, results vary from day 2 to day 5 by 58% and 95% [29].

The values of biodegradation of oil and fat obtained between 69% and 75%, and are similar to those of an aerobic biodegradation experiment that removes fats and oils from the dairy industry, which uses as biomass a mixture of isolated and selected native bacteria, which reached 72% efficiency of biodegradation [30]. The biodegradation of oils and greases incorporates the carbon of these substrates to the ecological chains and the carbon cycle, from the ecosystem point of view, is the most sustainable option.

5. Conclusions

- Sunflower oil is biodegradable and it is reached a percentage of biodegradation above 51% for a wide range of concentrations of these kinds of substrates in the influent, despite the competitive presence of saccharose.
- The removed oil in the process of wastewater treatment is through floating and biodegradation mechanisms. The elimination of sunflower oil sustained in biodegradation and flotation reaches 90%.
- It is concluded that it is pertinent to add superficial scavengers to the sedimentation tank that eliminate the floating oil.
- Biomass has a great capacity to adapt to an oil substrate and does not affect sedimentation.

Conflicts of Interest: The author declares no conflict of interest.

References

1. Dueholm, T.E.; Andreasen, K.H.; Nielsen, P.H. Transformation of lipids in activated sludge. *Water Sci. Technol.* **2001**, *43*, 165–172. [PubMed]
2. Chipasa, K.B.; Medrzycka, K. Behavior of lipids in biological wastewater treatment processes. *J. Ind. Microbiol. Biotechnol.* **2006**, *33*, 635–645. [CrossRef] [PubMed]
3. Sainz, J. Separación de aceites de efluentes industriales. *Chem. Eng.* **2004**, *409*, 93–99. (In Spanish)
4. De Turris, A.; Yabroudi, S.C.; Valbuena, B.; Cárdenas, C.; Herrera, L.; Rojas, C. Tratamiento de aguas de producción por flotación con aire disuelto. *Interciencia* **2011**, *36*, 211–218.
5. Chipasa, K.B.; Medrzycka, K. Characterization of the fate of lipids in activated sludge. *J. Environ. Sci.* **2008**, *20*, 536–542. [CrossRef]
6. Ronzano, E.; Dapena, J. *Tratamiento Biológico De Las Aguas Residuales*, 2nd ed.; Diaz de Santos: Madrid, Spain, 2002.
7. Rittmann, B.; McCarty, P. *Biotecnología del Medio Ambiente*; McGraw-Hill: Madrid, Spain, 2001.
8. Kurashige, J.; Matsuzaki, N.; Makabe, K. Modifications of fats and oils by lipases. *J. Dispers. Sci. Technol.* **1989**, *10*, 531–559. [CrossRef]
9. Ratledge, C. Microbial oxidations of fatty alcohols and fatty acids. *J. Chem. Technol. Biotechnol.* **1992**, *55*, 399–400. [CrossRef]
10. Sugimori, D.; Utsue, T. A study of the efficiency of edible oils degraded in alkaline conditions by *Pseudomonas aeruginosa* SS-219 and *Acinetobacter* sp. SS-192 bacteria isolated from Japanese soil. *J. Microbiol. Biotechnol.* **2012**, *28*, 841–848. [CrossRef] [PubMed]
11. Bitton, G. *Wastewater Microbiology*; John Wiley & Sons: New York, NY, USA, 1994.
12. Ramalho, R.S. *Tratamiento de Aguas Residuales*, 2nd ed.; Reverte Ed.: Saint-Armand, QC, Canada, 2003.
13. Volkering, F.; Breure, A.M.; Van Andel, J.G. Effect of micro-organisms on the bioavailability and biodegradation of crystalline naphthalene. *Appl. Microbiol. Biotechnol.* **1993**, *40*, 535–540. [CrossRef]
14. Jiménez, D.; Medina, S.A.; Gracida, J.N. Propiedades, aplicaciones y producción de biotensoactivos. *Int. J. Environ. Pollut.* **2010**, *26*, 65–84.
15. Weatherley, L.R.; Rooney, D.W.; Niekerk, M.V. Clean synthesis of fatty acids in an intensive Lipase-Catalysed Bioreactor. *J. Chem. Technol. Biotechnol.* **1997**, *68*, 437–441. [CrossRef]
16. Tsouris, C.; Neal, S.H.; Shah, V.M.; Spurrier, M.A.; Y Lee, M.K. Comparison of liquid-liquid dispersions formed by a stirred tank and elestrostatic spraying. *Chem. Eng. Commun.* **1997**, *160*, 175–197. [CrossRef]

17. Albasi, C.; Riba, J.P.; Sokolovska, O.; Bales, V. Enzimatic hydrolysis of sunflower oil, characterisation of interface. *J. Chem. Technol. Biotechnol.* **1997**, *69*, 329–336. [CrossRef]
18. Belitz, H.D.; Grosch, W. *Química De Los Alimentos*, 3rd ed.; Acribia, S.A., Ed.; Acribia Editorial: Zaragoza, Spain, 2009.
19. Crespi, M.; Huertas, J.A. Determinación simplificada de la demanda química de oxígeno por el método del dicromato. *Water Technol.* **1984**, *13*, 35–40.
20. APHA-AWWA-WPFC. *Métodos Normalizados Para el Análisis de Agua Potable y Aguas Residuales*; Diaz de Santos: Madrid, Spain, 1992.
21. Vittadini, G. *Catálogo de Información de Equipamiento de Biocontrol*; Vittadini Riferiment: Milan, Italy, 1991.
22. Eddy, M. *Wastewater Engineering, Treatment and Reuse*, 4th ed.; McGraw-Hill: Madrid, Spain, 2002.
23. Henze, M.; Harremoes, P.; Jansen, J.C.; Arvin, E. *Wastewater Treatment, Biological and Chemical Processes*; Springer: Berlin, Germany, 1995.
24. Cisterna, P.; Gutiérrez, A.; Sastre, H. Impact of previous acclimatization of biomass and alternative substrates in sunflower oil biodegradation. *Dyna* **2015**, *82*, 56–61. [CrossRef]
25. Cisterna, P.; Gutierrez, A.; Sastre, H. Biodegradación de aceite girasol con presencia de sacarosa mediante lodos activos a escala de laboratorio. *Interciencia* **2015**, *40*, 684–689.
26. Brooksbank, A.M.; Latchford, J.W.; Mudge, S.M. Degradation and modification of fats, oils and grease by commercial microbial supplements. *J. Microbiol. Biotechnol.* **2007**, *23*, 977–985. [CrossRef]
27. Young, J.C. Removal of grease and oil by biological treatment processes. *J. Water Pollut. Control Fed.* **1979**, *51*, 2071–2087. [PubMed]
28. Brenes, M.; Garcia, P.; Romero, C.; Garrido, A. Treatment of green table olive wastewater by activated-sludge process. *J. Chem. Technol. Biotechnol.* **2000**, *75*, 459–463. [CrossRef]
29. Wakelin, N.G.; Forster, C.F. An investigation into microbial removal of fats, oils and greases. *Bioresour. Technol.* **1997**, *59*, 37–43. [CrossRef]
30. Loperena, L.; Ferrari, M.D.; Díaz, A.L.; Guzmán, I.; Pérez, L.V.; Carvallo, F.; Travers, D.; Menes, R.J.; Lareo, C. Isolation and selection of native microorganisms for the aerobic treatment of simulated dairy wastewaters. *Bioresour. Technol.* **2009**, *100*, 1762–1766. [CrossRef] [PubMed]

![environments logo] *environments*

MDPI

Article

Geometric Factor as the Characteristics of the Three-Dimensional Process of Volume Changes of Heavy Soils

Milan Gomboš, Andrej Tall ⓘ **, Branislav Kandra, Lucia Balejčíková and Dana Pavelková** *

Institute of Hydrology, Slovak Academy of Sciences, Hollého 42, Michalovce 071 01, Slovakia;
gombos@uh.savba.sk (M.G.); tall@uh.savba.sk (A.T.); kandra@uh.savba.sk (B.K.); balejcikova@uh.savba.sk (L.B.)
* Correspondence: pavelkova@uh.savba.sk; Tel.: +421-56-251-47

Received: 24 January 2018; Accepted: 24 March 2018; Published: 27 March 2018

Abstract: During simulation of a water regime of heavy soils, it is necessary to know the isotropy parameters of any volume changes. Volume changes appear in both vertical and horizontal directions. In vertical directions, they appear as a topsoil movement, and in horizontal directions as the formation of a crack network. The ratio between horizontal and vertical change is described using the geometric factor, r_s. In the present paper, the distribution of volume changes to horizontal and vertical components is characterized by the geometric factor, in selected soil profiles, in the East Slovakian Lowland. In this work the effect of soil texture on the value of the geometric factor and thus, on the distribution of volume changes to vertical and horizontal components was studied. Within the hypothesis, the greatest influence of the clay soil component was shown by the geometric factor value. New information is obtained on the basis of field and laboratory measurements. Results will be used as inputs for numerical simulation of a water regime for heavy soils in the East Slovakian Lowland.

Keywords: volume changes; isotropy; geometric factor

1. Introduction

Heavy soils are characterized by a high content of clay particles which, due to moisture changes, cause volume changes in soil. When moisture increases, soil swells, and on drying, soil shrinks. Volumetric changes occur in a three-dimensional process, that in the vertical direction reflects the vertical movement of the soil surface, and in the horizontal plane the formation of cracks. The degree of distribution of soil volume changes to horizontal and vertical components depends on the isotropic properties of the soil environment [1–4]. The presence of cracks in conditions of heavy soils creates a two-domain soil environment, which significantly affects hydrological processes, especially in extreme meteorological conditions, highlighting the importance of studying this three-dimensional process [5–9]. Due to expected climatic changes, it is assumed that an increased frequency of extreme meteorological events will occur.

The aim of the paper is to characterize the distribution of volume changes to horizontal and vertical components in selected soil profiles in the East Slovakian Lowland (ESL) using a geometric factor. Based on the obtained results, the next goal is to verify the hypothesis concerning any effect of texture on the value of the geometric factor and thus, on the degree of distribution of volume changes to vertical and horizontal components. Within the hypothesis, it is assumed that the clay soil component has the greatest influence on the geometric factor value [10,11]. Results will be used as inputs for numerical simulation of water regimes of heavy soils in the ESL [12–17].

2. Materials and Methods

The procedure for determining isotropic soil volume expansion is based on laboratory determination of volume changes in clay-loam soil, for a precisely defined geometry of soil sample volume at the initial state of the experiment, and in the drying process under controlled conditions. Drying of isotropic soil samples is accompanied by volumes changes caused by horizontal and vertical shrinkage. Horizontal shrinking under natural conditions causes crack formations and vertical changes followed by decrease of soil surface area. From this point of view, laboratory results of studied volume changes should be interpreted for natural conditions. Previous studies [18–20] have mentioned one possible approach.

Figure 1 is a plot of soil sample saturation in the form of a cube with an edge length, z_s. Soil sample saturation is represented by a broken line. Its volume is $V_s = z_s^3$. After isotropic shake, the saturated sample is reduced by volume ΔV per cube, as per $z = z_s - \Delta z$ for edge length and volume $V = z^3$. Based on the above, it is possible to formulate a relationship

$$\frac{V}{V_s} = \frac{V_s - \Delta V}{V_s} = \frac{z^3}{z_s^3} = \frac{(z_s - \Delta z)^3}{z_s^3}. \tag{1}$$

On the basis of Equation (1) and Figure 1, it is possible to deduce the relationship between volume change of sample and change in sample height, i.e., by changing the volume in the vertical direction and in the horizontal plane.

$$1 - \frac{\Delta V}{V_s} = \left[1 - \frac{\Delta z}{z_s}\right]^3. \tag{2}$$

For most soils in natural conditions, according to Equation (2), it is possible to consider isotropic shrinkage. Therefore, it is possible to substitute a the geometric factor r_s for 3 without greater loss of accuracy. We can then present a generalized version of the equation in the form

$$1 - \frac{\Delta V}{V_s} = \left[1 - \frac{\Delta z}{z_s}\right]^{r_s}. \tag{3}$$

Mathematical expression of the relationship between volume change of sample and its vertical decrease can then be expressed in the form

$$\Delta z = z_s - \left[\left(\frac{V}{V_s}\right)^{\frac{1}{r_s}}\right] z_s. \tag{4}$$

Total change in soil volume is represented by $\Delta V = \Delta V_v + \Delta V_h$, while $\Delta V_v = z_s^2 \Delta z$. From the above and from Equation (4), the relation follows

$$\Delta V_h = \Delta V - z_s^2 \Delta z, \tag{5}$$

$$\Delta V_h = V_s \left[\left(\frac{V}{V_s}\right)^{\frac{1}{r_s}} - \frac{V}{V_s}\right]. \tag{6}$$

In accordance with Equation (1), Equation (2) can be interpreted as follows. In the case of shrinking without cracks, $r_s = 1$. In the case of crack formation and without decrease, $r_s \to \infty$. For all other cases, for formation of drained cracks and a surface decrease, occurring simultaneously, the following values are distinguished: $r_s = 3$ for isotropic shrinking; $1 < r_s < 3$, while vertical movement is more prevalent than horizontal shrinking. If $r_s > 3$, then the formation of cracks is more prevalent than vertical movement.

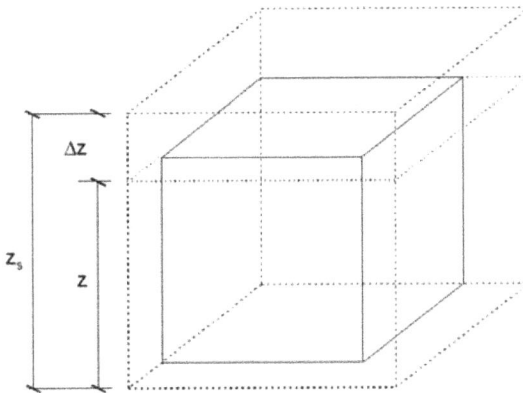

Figure 1. Change in the isotropic state of a cube shaped soil sample during drying. A dotted line shows the sample in saturated state, volume V_s, with an edge length z_s. The full line shows the sample after shrinkage, volume V, with an edge length z_s. Change of the cube edge length after drying is Δz.

Considering that during the drying, the change of sample height in the direction of the vertical axis of sampling is measured, and volume of soil sample saturated by water, and volume of dried soil sample is measured also, it is possible to calculate the geometric factor r_s from Equation (4). By adjusting the equation, we get a relationship for calculating the geometric factor in the following shape.

$$r_s = \frac{\log\left(\frac{V}{V_s}\right)}{\log\left[\frac{-(\Delta z)+z_s}{z_s}\right]}. \tag{7}$$

From Equation (5), based on these theoretical considerations and experimental field and laboratory measurements, the geometric factor of volume expansion was determined as a quantifier of the distribution of total volume changes to horizontal and vertical components.

Another part of this research consists of two stages. In the first stage, r_s values were quantified for characteristic profiles on the ESL. In the second stage, the correlation between values of the geometric factor and content of clay ($r_s < 0.002$ mm), dust ($r_s = 0.002$–0.050 mm) and sandy particles ($r_s = 0.050$–2.000 mm) in soil were investigated. Investigated soil layers were classified in terms of isotropic volume changes.

To assess this correlation, models that best describe the tightness of correlation relationships were selected. As a criterion for selecting a model, the correlation coefficient value was used.

In addition, the relationship between r_s and V_h has been shown, based on measurements of V, V_s, Z_s, Δz and Equation (6).

Measurements were made on the ESL in 11 sampling profiles from 10 different locations, the location of which is shown in Figure 2. Sampling points were selected to characterize the soil environment of the ESL. As part of the fieldwork, intact soil samples were taken by the swab probe method and placed into collecting rollers. The volume of withdrawal rollers was 100 cm^3 with a diameter of 5.6 cm. The geometric dimensions of soil samples during shaking were measured by a micrometer. The size of shrinking versus percentage saturation for Michalovce is shown in Table 1, in the last column on the right. The largest measured volume changes were around 40%. Soil samples were saturated under laboratory conditions to full water capacity, and after measuring their geometric dimensions, were dried at 30 °C gradually until the weight changes were approaching the measurement error. The last drying was carried out at 105 °C. All necessary data for calculating the variables of Equation (7) were obtained by measuring the geometric dimensions of the soil samples after drying.

Soil samples were taken at depths of 0.10 m to 0.80 m at all locations except the Milhostov site. At the Milhostov site, samples were taken from depths of soil up to 1.60 m.

Figure 2. Map of the area of interest.

3. Results

The results of the measurements and calculations were divided into two parts in accordance with the methodological procedure. In the first part, geometric factor values were quantified. In the Michalovce profile in Table 1, the values of the soil sample volume changes, and values of the geometric factor r_s, are shown. In Table 1, for three successive layers of 0.1 m intervals, the height of the water-saturated rollers, their saturated volumes and these values after drying, are given. For each measurement, a calculated value of r_s and then an average value of the geometric factor for each layer are calculated, according to Equation (7). From Table 1, percentage volume change, and change in vertical dimension (Δz) of individual samples, as well as their average values in each layer, are obvious. In Table 2, each layer of each collection site is characterized by texture composition. All eleven sites were processed analogously. Table 3 summarizes values of the geometric factor in all eleven investigated soil profiles. The table also shows coordinates of individual field profiles. Results indicate that, in the examined profiles, volume change predominantly results in the formation of cracks ($r_s > 3.0$). Only in Senné was an isotropic shrinkage identified ($r_s = 3.0$). In two profiles, the predominance of vertical movements over crack formation was identified (Kamenec (2.8), Somotor 2 (2.3)). The highest value ($r_s = 13.0$) was identified in the locality of Vysoká [21]. Vysoká also had the highest variation range. The lowest value ($r_s = 1.7$) was found in Somotor 2.

The second part of this research focused on a correlation analysis between the values of the geometric factor and the components of the texture of the soil samples. In addition, the shape of dependence between r_s and V_h is shown based on volumetric measurements.

Figure 3 represents an analysis of the effect of the clay particle fraction, which is represented by a particle fraction of <0.002 mm for the geometric factor values. The geometric factor in this case is the average value of soil layers from which soil samples were taken for grain analysis. The effect of the content of clay particles on r_s is obvious from this picture. In this case a correlation was identified $R = 0.56$. The result can be interpreted as a significant influence of clay content in soil on the geometric factor, and on the distribution of total volume change to vertical and horizontal components. The figure shows that clay, and clay soil (45% to 55% clay content), are isotropic in terms of volume change. In addition, vertical movement slightly outweighs horizontal plane movement with an increase in the clay content of soils, and the variability of this movement decreases. The geometric factor oscillates slightly below the isotropic boundary. As the proportion of clay in soils decreases, the variability of r_s is significantly increased.

Table 1. Soil sample measurements and calculation of r_s in the Michalovce locality.

Layer [cm]	Rolle Number	z_s [m]	V_s [m³]	V [m³]	Z [m]	Δz [m]	r_s [-]	$\overline{r_s}$ [-]	Δz [%]	ΔV [%]
	88	4.38	101.02	80.44	4.06	0.32	2.98		7.36	20.38
0–10	507	4.30	100.97	81.58	4.07	0.30	3.04	2.93	6.77	19.20
	511	4.33	101.25	84.29	4.11	0.28	2.76		6.43	16.75
	78	4.30	100.87	85.90	4.16	0.20	3.41		4.59	14.80
10–20	527	4.30	100.31	86.11	4.15	0.22	2.96	3.05	5.03	14.16
	530	4.33	100.51	86.28	4.13	0.23	2.79		5.33	14.16
	504	4.30	100.04	87.84	4.16	0.19	2.91		4.37	12.20
20–30	525	4.33	100.51	88.92	4.18	0.18	2.94	3.09	4.08	11.53
	533	4.37	100.85	88.79	4.21	0.16	3.40		3.67	11.96
	74	4.36	100.63	87.16	4.16	0.20	3.10		4.53	13.38
30–40	87	4.37	101.01	87.19	4.21	0.16	3.94	3.36	3.66	13.69
	515	4.34	100.42	85.93	4.13	0.22	3.03		5.01	14.43
	76	4.36	100.83	90.10	4.26	0.11	4.59		2.41	10.64
40–50	80	4.37	101.15	90.61	4.28	0.10	5.00	4.58	2.17	10.42
	519	4.38	99.70	90.35	4.28	0.10	4.15		2.34	9.37
	84	4.37	100.97	90.57	4.24	0.13	3.60		2.97	10.30
50–60	506	4.35	100.41	91.39	4.25	0.10	4.04	4.07	2.30	8.98
	536	4.36	99.79	92.12	4.28	0.08	4.57		1.72	7.68
	81	4.36	100.43	93.63	4.33	0.03	-		0.69	6.78
60–70	529	4.36	100.13	92.82	4.30	0.06	5.45	5.27	1.38	7.30
	535	4.37	101.05	93.35	4.30	0.07	5.08		1.55	7.62
	85	4.38	100.96	95.62	4.33	0.05	4.70		1.14	5.29
70–80	505	4.36	100.79	93.73	4.31	0.06	5.46	4.76	1.32	7.01
	534	4.35	100.39	93.45	4.28	0.08	4.12		1.72	6.91

Legend: z_s—sample height in a saturated state; V_s—volume of the saturated sample; V—sample volume after shrinkage; Z—height after drying, ΔZ—change of sample height; r_s—geometric factor; $\overline{r_s}$layer average of the geometric factor; ΔV—total change in sample volume.

Table 2. Texture characteristics of studied soil profiles in Michalovce. Analogously, all examined profiles were characterized.

Layer [cm]	I. Fraction <0.001 mm	Clay [%]	Silt [%]	Sand [%]
0–10	28.07	31.85	56.08	12.07
10–20	27.19	31.04	59.27	9.69
20–30	27.29	31.14	58.42	10.44
30–40	27.61	31.63	59.89	8.48
40–50	23.09	26.66	60.83	12.51
50–60	19.78	22.73	60.67	16.60
60–70	18.35	20.89	63.53	15.59
70–80	16.94	19.27	56.33	24.40

Table 3. Profile evaluation by the average r_s value.

Locality [cm]	Coordinates	r_s in Layer [-]				I. Fraction <0.001 mm	Clay <0.002 mm	Profile Rating
		Max	Min	Max-Min	Avg			
Michalovce	N48°44.255′ E21°56.664′	5.3	2.9	2.3	3.9	23.54	26.90	1
Milhostov	N48°40.185′ E21°44.248′	5.9	2.8	3.1	3.9	27.28	29.15	1
Pribeník	N48°23.688′ E21°59.547′	3.4	2.9	0.5	3.1	29.70	31.90	1
Senné	N48°39.802′ E22°02.892′	4.4	2.6	1.8	3.0	51.03	54.84	2
Sírnik	N48°30.538′ E21°48.830′	4.8	2.7	2.0	3.3	30.55	35.18	1
Somotor 1	N48°23.748′ E21°48.471′	10.0	3.8	6.2	5.3	23.72	25.89	1
Somotor 2	N48°23.173′ E21°48.237′	3.0	1.7	1.3	2.3	18.30	20.15	3
Horeš	N48°22.540′ E21°53.907′	3.9	2.7	1.2	3.1	41.19	43.74	1
Kamenec	N48°21.048′ E21°48.877′	3.4	2.2	1.2	2.8	29.81	32.11	3
Vysoká	N48°36.796′ E22°06.898′	13.0	3.0	10.0	5.7	11.27	12.88	1
Zatín	N48°28.725′ E21°54.918′	3.4	2.9	0.5	3.4	28.45	31.55	1

Legend: 1—formation of cracks ($r_s > 3$); 2—isotropic shrinkage ($r_s = 3$); 3—vertical movement $1 < r_s < 3$.

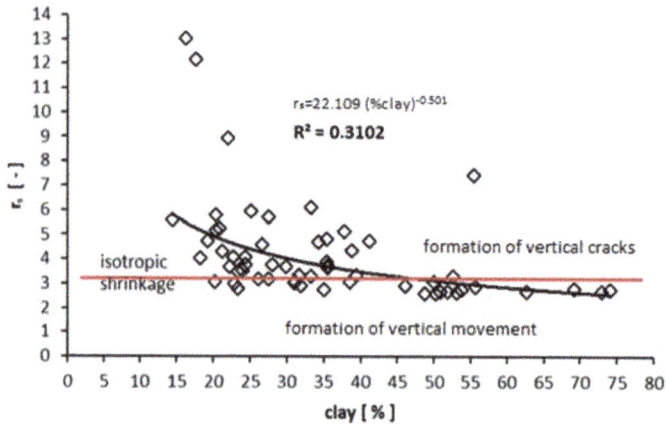

Figure 3. Relationship between the geometric factor, r_s, and grain content of particles sized <0.002 mm, which represents the percentage of clay particles in soil samples.

Figure 4 shows the relationship between r_s and silt particles in the soil samples. The Correlation coefficient, $R = 0.34$, is insignificant. The greatest variability of r_s was found to be between 30 and 40% for silt particles. Increasing the particle ratio caused the variability of r_s to decrease, and r_s values tended to approach the value of "3" from above, i.e., to the value of isotropic volume change.

Similar results are documented in Figure 5. Here, the relationship between r_s and the sand content in soil can be seen. The correlation coefficient of $R = 0.36$ is insignificant. The difference between the sand content in soil and silt content in soil, is the evenly distribution of sand content variability. The sand content trend in soil, increasing with the geometric factor, is opposite to that of the silt content in soil, which decreases with an increase in the geometric factor. In both cases, the r_s values oscillate around the isotropic shrinkage value, or are higher than the isotropic shrinkage value.

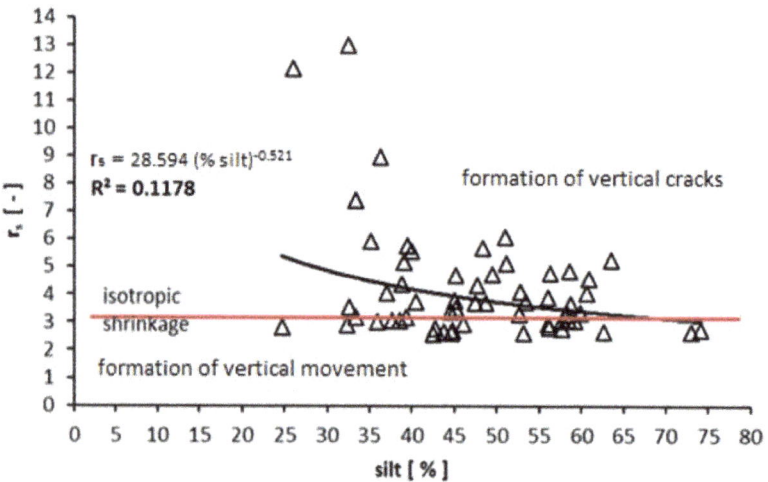

Figure 4. Relationship between the geometric factor, r_s, and the grain content of particles sized between 0.002–0.05 mm, which represents the percentage of silt particles in the soil samples.

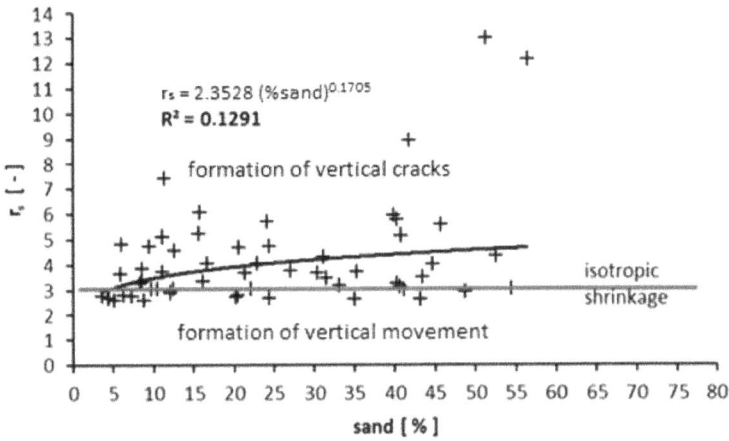

Figure 5. Relationship between the geometric factor, r_s, and the grain content of particles sized between 0.05–2 mm, which represents the percentage of sand particles in the soil samples.

Figure 6 shows the relationship between the volume change in the horizontal direction V_h, and the geometric factor, r_s, with the volume change V_h being expressed as a percentage of the total measured volume change. The graph shows the effect of the geometric factor on the behavior of volumetric changes and their distribution to horizontal and vertical components. With an increase in r_s, more of the total volume change is due to the horizontal component.

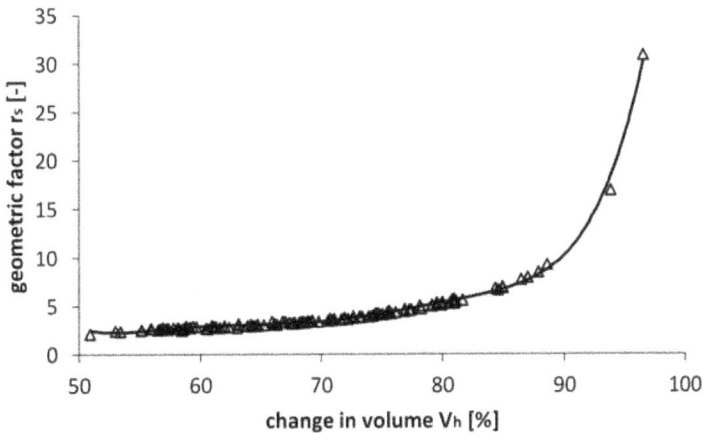

Figure 6. Relationship between the geometric factor and the horizontal component of volume change.

4. Discussion

There may be several causes of the observed anisotropy of soil, when its volume was changed. It can be assumed that this is due to inequalities that occur in sedimentation processes, and the inhomogeneity of clay minerals that are sources of volume changes in soils. The primary condition causing volume change, is the presence of clay minerals and water. Crystals of clay mineral are composed of platelet formations (consisting of silicon tetrahedra and aluminum octahedra) ranging from 1 (montmorillonite) to a theoretically unlimited number (kaolinite). Individual plates have a very small valency (0.5–1.0 nm) and a high specific surface area (15 m^2·g^{-1}—kaolinite, 80 m^2·g^{-1}—illite; 800 m^2·g^{-1}—montmorillonite). The size of the specific surface of these plates is closely related to the size of the volume changes. The larger the specific surface, the greater the ability to swell. The surface of the clay mineral plates carries a negative electrical charge that can attract water molecules. Consequently, with the increasing the surface area of clay mineral crystals, the more water molecules can bind. With an increasing water content, swelling of such a clay occurs. Drying will return the clay to its original volume. The abilities of individual minerals to assimilate water into their structure, and thus increase their volume, are different. Kaolinite clays are relatively inactive (they have low ability to bind water), illite clays have low to moderate swelling ability, and montmorillonite clays are highly expansive, and in their pure form (under laboratory conditions), can change volume by 1400–2000% (Na-montmorillonite). Pore size and pore distribution in soil is another factor influencing volume change. Presence of organic components in soil may also have a significant effect on volume changes.

Under natural conditions, clay minerals do not occur in pure form but in mixed structures. Therefore, it is necessary to know the species composition of clay minerals for a detailed assessment of these processes. Their determination is difficult. Information on the species composition of clay minerals is insufficient for the ESL.

For these reasons, further research on the ESL should aim to determine the species composition of clay minerals in characteristic soil species. These findings will help to avoid possible mistakes in determining cracked porosity based on physical clay content. This is the reason for paying increased attention to identifying the species composition of clay minerals in the soils of the ESL.

In this context, it is necessary to be aware that the amount and distribution of volumetric changes to horizontal and vertical components also have an indirect influence on the presence of silt and sand in soil. The representation of any of the textural components of clay, silt and sand, is a linear combination of the remaining two, provided that the granularity is accurately determined. If one of them is increased in soil, the remaining two are reduced. Silt and sand do not cause volume changes, but the increased soil content reduces clay content and volume change. This creates an inhomogeneous environment in terms of volume change and the associated geometric factors. The anisotropy of volume changes increases followed by the increased variability of the geometric factor. If there are large amounts of silt and sand or a reduction in clay, then soil is lighter and volume changes disappear, respectively. These changes are not measurable. With an increase in clay content and a decrease in silt and sand content, volume changes begin to display characteristics of isotropic shrinkage, i.e., $r_s = 3$. With isotropic shrinkage, the impacts of silt and sand gradually disappear. Changes in r_s across the vertical soil profile may be related to an increase or decrease in soil clay content. This is common under the ESL conditions. This may be associated with the impact of superficial soil layers. Sedimentation processes can have a significant impact on the anisotropy of volumetric changes and changes in the geometric factor, r_s, across vertical soil profiles in lowland conditions. This is especially true when platelet clay particles are predominately oriented in one direction.

The degree of influence of individual grain fractions on the geometric factor is related to the size of the grains and the pore size. The smaller the particles in the porous material, the more homogeneous it is, and the value r_s approximates or oscillates around the value "3". The upper limit of clay particle size (2 μm) is 25 times smaller than the upper limit of dust particle size (50 μm) and 1000 times less than the upper limit of sand particle size (2000 μm). These variation ranges cause r_s to deviate from the value "3", i.e., to deviate from isotropic shrinkage and thus increases the variability of r_s.

5. Conclusions

In this paper, values of the geometric factor, r_s, were quantified in characteristic soil profiles. Based on this, soil profiles were evaluated for the distribution of volume changes to horizontal and vertical components. In one case, isotropic volume changes were identified in Senné. This was the locality with the highest content of clay minerals of all investigated localities. Vertical volume changes dominated in two locations. In the remaining eight locations, horizontal volume changes dominated, i.e., crack formation and thus the formation of a two-domain soil system (crack and soil matrix) [22–25]. Summaries of the results are shown in Table 3.

The effect of texture on the geometric factor was analyzed. A high correlation between measured horizontal volume changes and the geometric factor has been demonstrated. The results are graphically shown in Figures 3–6. All results were obtained from experimental measurements in the field and in the laboratory.

To focus further research on this subject area, we recommend paying increased attention to identifying the species composition of clay minerals in the soils of the ESL.

The results of this analysis will be used for numerical simulation of the water regime and its prognosis under heavy soil conditions in the ESL.

Acknowledgments: This study is the result of the project implementation: Centre of excellence for the integrated river basin management in the changing environmental conditions, ITMS code 26220120062; supported by the Research & Development Operational Programme funded by the ERDF. Authors are also grateful to Scientific Grant Agency of the Ministry of Education, Science, Research and Sport of the Slovak Republic: (project VEGA: 2/0062/16).

Author Contributions: Milan Gomboš Proposed the methodology and conceived the of experiments, interpreted the results, and wrote the article; Andrej Tall and Branislav Kandra performed field work and experimental laboratory work and analyzed the data; Lucia Balejčíková processed experimental data and prepared sections of the text; Dana Pavelková performed laboratory work, data analysis and prepared sections of the text.

Conflicts of Interest: The authors declare no conflict of interest.

References

1. Boivin, P. Anisotropy, cracking, and shrinking of vertisol samples. Experimental study and shrinkage modeling. *Geoderma* **2007**, *138*, 25–38. [CrossRef]
2. Cornelis, W.M.; Corluy, J.; Medina, H.; Diaz, J.; Hartmann, R.; Van Meirvenne, M.; Ruiz, M.E. Measuring and modeling the soil shrinkage characteristic curve. *Geoderma* **2006**, *137*, 179–191. [CrossRef]
3. Chertkov, V.Y. An integrated approach to soil structure, shrinkage, and cracking in samples and layers. *Geoderma* **2012**, *173–174*, 258–273. [CrossRef]
4. Tall, A. Porovnanie klasifikačných systémov pre určovanie textúry pôd so zameraním na ťažké pôdy. *Acta Hydrol. Slov.* **2002**, *3*, 87–93.
5. Krisnanto, S.; Rahardjo, H.; Fredlund, D.G.; Leong, E.C. Water content of soil matrix during lateral water flow through cracked soil. *Eng. Geol.* **2016**, *210*, 168–179. [CrossRef]
6. Laufer, D.; Loibl, B.; Marlander, B.; Koch, H.J. Soil erosion and surface runoff under strip tillage for sugar beet (*Beta vulgaris* L.) in Central Europe. *Soil Tillage Res.* **2016**, *162*, 1–7. [CrossRef]
7. Lopez-Bellido, R.J.; Munoz-Romero, V.; Lopez-Bellido, F.J.; Guzman, C.; Lopez-Bellido, L. Crack formation in a mediterranean rainfed Vertisol: Effects of tillage and crop rotation. *Geoderma* **2016**, *281*, 127–132. [CrossRef]
8. Novák, V. Soil-crack characteristics—Estimation methods applied to heavy soils in the NOPEX area. *Agric. For. Meteorol.* **1999**, *98–99*, 501–507. [CrossRef]
9. Novák, V.; Šimunek, J.; Van Genuchten, M.T. Infiltration of water into soil with cracks. *J. Irrig. Drain. Eng.* **2000**, *126*, 41–47. [CrossRef]
10. Chertkov, V.Y. The shrinkage geometry factor of a soil layer. *Soil Sci. Soc. Am. J.* **2005**, *69*, 1671–1683. [CrossRef]
11. Chertkov, V.Y.; Ravina, I.; Zadoenko, V. An approach for estimating the shrinkage geometry factor at a moisture content. *Soil Sci. Soc. Am. J.* **2004**, *68*, 1807–1817. [CrossRef]
12. Koltai, G.; Hegedős Mikéné, F.; Végh, K.R.; Orfánus, T.; Rajkai, K. Soil moisture monitoring as resilience indicator on the Danube lowlands. *Növénytermelés* **2010**, *59*, 291–294.
13. Saadeldin, R.; Henni, A. A novel modeling approach for the simulation of soil-water interaction in a highly plastic clay. *Geomech. Geophys. Geo-Energy Geo-Resour.* **2016**, *2*, 77–95. [CrossRef]
14. Šoltész, A.; Baroková, D. Impact of landscape and water management in Slovak part of the Medzibodrožie region on groundwater level regime. *J. Landsc. Manag.* **2011**, *2*, 41–45.
15. Števulová, N.; Balintová, M.; Zeleňáková, M.; Eštoková, A.; Vilčeková, S. Environmental Engineering in the Slovak Republic. *IOP Conf. Ser. Earth Environ. Sci.* **2017**, *92*, 012064. [CrossRef]

16. Tarnik, A.; Igaz, D. Quantification of soil water storage available to plants in the Nitra river basin. *Acta Scientiarum Polonorum-Formatio Circumiectus* **2015**, *14*, 209–216. [CrossRef]

17. Velísková, Y. Changes of water resources and soils as components of agro-ecosystem in Slovakia. *Növénytermelés* **2010**, *59*, 203–206.

18. Bronswijk, J.J.B.; Evers-Vermeer, J.J. Shrinkage of Dutch clay soil aggregates. *Neth. J. Agric. Sci.* **1990**, *38*, 175–194.

19. Bronswijk, J.J.B. Prediction of actual cracking and subsidence of clay soils. *Soil Sci.* **1989**, *148*, 87–93. [CrossRef]

20. Bronswijk, J.J.B. Relation between vertical soil movement and water—Content changes in cracking clays. *Soil Sci. Soc. Am. J.* **1991**, *55*, 1220–1226. [CrossRef]

21. Peng, X.; Horn, R. Anisotropic shrinkage and swelling of some organic and inorganic soils. *Eur. J. Soil Sci.* **2007**, *58*, 98–107. [CrossRef]

22. Coppola, A.; Gerke, H.H.; Comegna, A.; Basile, A.; Comegna, V. Dual-permeability model for flow in shrinking soil with dominant horizontal deformation. *Water Resour. Res.* **2012**, *48*, W08527. [CrossRef]

23. Chertkov, V.Y. The geometry of soil crack networks. *Open Hydrol. J.* **2008**, *2*, 34–48. [CrossRef]

24. Xing, X.; Liu, Y.; Ma, X. Effects of soil additive on soil-water characteristic curve and soil shrinkage. *Shuikexue Jinzhan/Adv. Water Sci.* **2016**, *27*, 40–48.

25. Zhu, L.; Chen, J.; Liu, D. Morphological quantity analysis of soil surface shrinkage crack and its numerical simulation. *Nongye Gongcheng Xuebao/Trans. Chin. Soc. Agric. Eng.* **2016**, *32*, 8–14. [CrossRef]

environments

MDPI

Article

Prediction of Reservoir Sediment Quality Based on Erosion Processes in Watershed Using Mathematical Modelling

Natalia Junakova [1,*], Magdalena Balintova [1], Roman Vodička [2] and Jozef Junak [1]

[1]   Faculty of Civil Engineering, Institute of Environmental Engineering, Technical University of Košice,
     042 00 Košice, Slovakia; magdalena.balintova@tuke.sk (M.B.); jozef.junak@tuke.sk (J.J.)
[2]   Faculty of Civil Engineering, Institute of Construction Technology and Management,
     Technical University of Košice, 042 00 Košice, Slovakia; roman.vodicka@tuke.sk
*   Correspondence: natalia.junakova@tuke.sk; Tel.: +421-55-602-4266

Received: 30 November 2017; Accepted: 27 December 2017; Published: 29 December 2017

Abstract: Soil erosion, as a significant contributor to nonpoint-source pollution, is ranked top of sediment sources, pollutants attached to sediment, and pollutants in the solution in surface water. This paper is focused on the design of mathematical model intended to predict the total content of nitrogen (N), phosphorus (P), and potassium (K) in bottom sediments in small water reservoirs depending on water erosion processes, together with its application and validation in small agricultural watershed of the Tisovec River, Slovakia. The designed model takes into account the calculation of total N, P, and K content adsorbed on detached and transported soil particles, which consists of supplementing the soil loss calculation with a determination of the average nutrient content in topsoils. The dissolved forms of these elements are neglected in this model. Validation of the model was carried out by statistical assessment of calculated concentrations and measured concentrations in Kľušov, a small water reservoir (Slovakia), using the *t*-test and *F*-test, at a 0.05 significance level. Calculated concentrations of total N, P, and K in reservoir sediments were in the range from 0.188 to 0.236 for total N, from 0.065 to 0.078 for total P, and from 1.94 to 2.47 for total K. Measured nutrient concentrations in composite sediment samples ranged from 0.16 to 0.26% for total N, from 0.049 to 0.113% for total P, and from 1.71 to 2.42% for total K. The statistical assessment indicates the applicability of the model in predicting the reservoir's sediment quality detached through erosion processes in the catchment.

Keywords: agricultural watershed; nonpoint-source pollution; nutrient content; sediment quality

---

1. Introduction

Soil erosion, as a significant contributor to nonpoint-source pollution, is ranked the top of sediment source [1], pollutants attached to sediment [2], and pollutants in the solution of surface water [3].

Worldwide, soil erosion by water affects 1094 million hectares of arable land [4]. Across Europe, data on trends in soil erosion are lacking and erosion estimates are based on modelling studies. In the 1990s, water erosion affected 105 million hectares of soil or 16% of Europe's total land area (excluding Russia) [5]. In 2006, it was estimated that the surface area affected by water erosion in the EU-27 was 130 million hectares [6]. In 2014–2015, approximately 11.4% of the EU territory was affected by moderate to high level water erosion rate (more than five tons per hectare per year). The reduction of this rate against 1990s by 4.6% is mainly due to the application of water erosion control practices which have been applied during the last decade in the EU [7].

A significant amount of global sediment flux is retained in reservoirs [8]. It is estimated that the global annual loss in storage capacity of the world's reservoirs due to sediment deposition is approximately 0.5–1% [9], and for individual reservoirs these values can be as high as 4–5% [10]. The useful lifetime of the reservoirs is thus reduced to only 22 years on average [11]. Other studies found in the literature [12,13] have reported that worldwide rivers carry approximately 15 billion tons of sediments to the sea annually. Walling and Webb [14] have given an overview of mean annual total suspended sediment transport to the oceans that ranged from 8 to 51 billion tons of sediments.

Together with the small fraction of sediment, pollutants including nutrients are transported via surface runoff [15] from arable land in the catchment [16] and are deposited in reservoirs [17]. The quantity of nutrient concentrations in water reservoir sediments, associated with nonpoint source pollution from agricultural catchments, is regarded as the environmental pollution index. As published by Qian et al. [18] the easier transformation of nitrogen and phosphorus from agricultural soils to freshwater bodies contributes to their accelerated eutrophication. Nutrient concentrations in runoff are affected by many factors including climate, soil characteristics, relief, land use, and chemical application [19]. Typically, the eroded soils contain about three times more nutrients per unit weight than are left in the remaining soil [20]. Pimentel and Burgess [21] summarized, that a ton of fertile topsoil averages 1 to 6 kg of nitrogen, 1 to 3 kg of phosphorus, and 2 to 30 kg of potassium, whereas the topsoil on the eroded land has an average nitrogen content of only 0.1 to 0.5 kg per ton. As reported in literature [22,23], sediments act as an efficient trap for both nutrients (nitrogen and phosphorus). For example, sediments from Gulf of Finland trap 20–50% of P and 40–65% of N in the Neva estuary and in the open Gulf, and up to 100% in the Neva Bay are buried within the accumulating sediment. The rest is released to the overlying water [22]. Similarly, [23] showed that finer sediment particles (silt and clay) transported by rivers carry the major part of nutrient loads by absorption and thus, sediment settling can remove nutrients from the water column. On the other hand, the sediment accumulation in Iron Gate reservoir (Romania) on the Danube River corresponded to 5% of total nitrogen and 12% of total phosphorus of the incoming loading [24].

In order to protect surface water resources and optimize their use, soil and nutrient losses from catchment areas must be controlled [25] and minimized [26]. Conventional methods to assess soil erosion [27] and sediment-associated chemical runoff are expensive, time-consuming, and need to be collected over many years [28,29]. Nowadays, preference is given to predict [30] and control sediment and nutrient yields from agricultural nonpoint source runoff using mathematical models [31]. Soil erosion models can assess and simulate the extent and magnitude of erosion processes in watershed. To predict soil erosion rates by water, several models exist which differ greatly in terms of complexity, inputs, and spatial and temporal scale [32]. Most have been developed for large agricultural areas and are designed to predict annual rates of soil loss from land under various cropland and rangeland management techniques [33,34]. To estimate sediment yields into small reservoirs, regression models can also be used. Most of these models are site-specific and do not permit the land and water managers to assess the impact of agronomic and mechanical changes on sediment yields [35] much less on sediment quality. Thus, there is a pressing need to extend such models to provide sediment quality prediction by integrating already available research methods with new generalization and integration techniques. Better predicting of small water reservoirs' sediment quantity and quality is necessary, of all things, with regard to utilization or application of dredged sediment from water reservoirs [36,37].

The objective of this study is to design a mathematical model intended to predict the total content of nitrogen, phosphorus, and potassium in bottom sediments in small water reservoirs, with its application and validation in the small agricultural watershed of the Tisovec River, Slovakia. Specification of sediment quality in the reservoir depending on the main periods of the cropping cycle and distribution of erosivity during a year is the novelty of this article. Also, an average plant nutrient uptake for chosen crops divided into five crop-stage periods during growing season was devised.

2. Materials and Methods

2.1. Study Area

A prediction model was designed on the basis of the nutrient transport study carried out in the small agricultural catchment of the Tisovec River (northeast of Slovakia, the district of Bardejov) covering an area of 6.0 km². The catchment falls into the drainage basin of the Topl'a River. The average annual air temperature is 8 °C and the average annual rainfall is about 670 mm with a maximum in the summer months. Planosols, cambisols, and albic luvisols are the dominant major soil groups of the Tisovec basin with the presence of medium-textured soils (sandy loam). Non-point sources of pollution from agricultural production areas are the leading cause of the sediment and water quality degradation in this catchment. Different land uses occur within the catchment.

The upper and middle zone of the catchment consists mainly of forests (39.2%) and pastures (21.7%), 21.4% of downstream areas are covered with agricultural land. The rest of the catchment is for other uses.

For irrigation of surrounding agricultural land and accumulation of water, the Kl'ušov small water reservoir with total capacity of 72,188 m³ was built in the Tisovec River in 1986. At present, it is also used for retention of high water, suburban recreation, and as a fishery. It has a surface area of 2.2 ha, a length of about 494 m, and a mean depth of 3.5 m. At the dam, the reservoir reaches a maximum water depth of 9.57 m [38].

Because of the rough terrain, climatic conditions, and soil types, this catchment is exposed to water erosion. Eroded soil particles in this catchment greatly affect the quantity and quality of sediments in the Kl'ušov reservoir. This reservoir trapped approximately 24,500 m³ of sediments delivered from the upper catchment during 19 years of its operation and its total storage capacity decreased by about 33% [38].

2.2. Sampling Procedure and Chemical Analysis

To determine the total N, P, and K content in eroded soil particles, soil samples were collected from two parcels (1004/1 and 2001/1) of arable land situated next to the reservoir (Figure 1). Within each parcel, about 30 to 40 soil subsamples were taken and mixed into one composite sample. Considering the fact that the highest nutrient levels occur in the surface layers, sampling depth was set at 0.30 m [39].

S1	49° 15′ 3.6241546″ N 21° 14′ 2.7187443″ E	S7	49° 15′ 2.9308355″ N 21° 13′ 58.9915466″ E
S2	49° 15′ 3.7376066″ N 21° 14′ 1.7338371″ E	S8	49° 15′ 2.212302″ N 21° 13′ 59.6095276″ E
S3	49° 15′ 3.3216157″ N 21° 14′ 2.0042038″ E	S9	49° 15′ 1.0777694″ N 21° 13′ 57.91008″ E
S4	49° 15′ 4.9351506″ N 21° 13′ 59.1074181″ E	S10-S14	49° 15′ 3.485491″ N 21° 14′ 1.2896633″ E
S5	49° 15′ 4.2418366″ N 21° 13′ 58.9915466″ E	S15-S22	49° 15′ 3.7754239″ N 21° 13′ 58.7984276″ E
S6	49° 15′ 3.2081635″ N 21° 14′ 0.942049″ E		

Figure 1. Particle size distribution measured for original and mechanically activated sediments.

Together with soil sampling, 21 composite (disturbed) reservoir sediment samples were taken from the drained reservoir to determine total N, P, and K content. Sampling methodology was as follows. Part of the composite sediment samples were taken from a location in proximity to the dam due to deposition of fine-grained particles (<63 μm) washed away through water erosion and preferentially attaching the nutrients [40]. Other samples were collected along the reservoir and others in different sampling depths from one location. A S9 sample was taken from the Tisovec stream.

Composite soil and sediment samples were collected in buckets; the weight of the composite samples was about 3 kg for sediment and 5 kg for soil. In laboratory conditions, the samples were air dried at room temperature, any coarse lumps were crushed and samples were homogenized.

Adsorbed forms of nutrients in soil and sediment samples were determined for total nitrogen content by elemental analysis (LECO CHN628) and for total phosphorus and potassium using an inductively coupled plasma-atomic emission spectrometry technique (Agilent 5100 ICP-OES).

To determine the amount of dissolved N and P in soil samples, laboratory leaching experiments were conducted. Leachates were prepared in a 1 to 10 proportion of soil sample to distilled water. After 24 h, leachates were subsequently filtered and total N and P contents were determined using a DR 890 (Hach Lange) portable colorimeter.

2.3. Model

The model works on the calculation of total nutrient (N, P, and K) concentrations in detached eroded soil particles in dissolved and adsorbed form of these elements.

Calculation of total N, P, and K in adsorbed form consists of supplementing the soil loss calculation with a determination of the average nutrient content in topsoils.

Soil loss from arable land is computed using the Universal Soil Loss Equation (USLE) [41] which is expressed as

$$G_{r,i} = R * K * L * S * C_{r,i} * P \tag{1}$$

where $G_{r,i}$ is the potential long term average annual soil loss (tons per hectare per year); R is the rainfall erosivity factor in MJ ha^{-1} cm h^{-1}; K is the soil erodibility factor in t ha h ha^{-1} MJ^{-1} mm^{-1}; L, S is the topographical factor; $C_{r,i}$ is the average plant cover factor in *i*-year calculated as a sum of divided C_i factors; and P is the support practice factor. The generalized plant cover (C) factor values resulting from the USLE are specified and modified by their dividing into five crop-stage periods (C_i) according to Wischmeier and Smith [41] (seedbed preparation, establishment, development, maturing crop, stubble field) to take into account the height of plant cover from the ground [42] and annual rainfall distribution. To compute $C_{r,i}$, partial C_i factor is weighted according to distribution of erosivity during a year.

The calculation of average nutrient concentrations in topsoils is expressed by deducting nutrient input from fertilizer use from plant nutrient uptake (output). As in soil loss calculation, even in this case, the calculation is divided into five crop-stage periods.

Finally, the total N, P, and K content in transported soil particles from arable land to reservoir is calculated through the modification of the average N, P, and K concentrations in eroded soil particles in adsorbed form detached through water erosion by sediment enrichment ratio (*SER*) [43] using the proposed equation

$$C_X = \frac{\sum_{i=1}^{5} X_i \cdot G_i}{G_{r,i}} \cdot SER = \frac{\sum_{i=1}^{5} X_i \cdot G_i}{G_{r,i}} \cdot e^{2-0.2\ln(G_{r,i}\cdot 1000)} \tag{2}$$

$$X_i = background\,soil\,concentration + (N, P, K\,input - plant\,N, P, K\,uptake) \tag{3}$$

where C_X is an average annual concentration of total N, P, K in transported soil particles from the studied parcel (kg N,P,K ha^{-1} or mg N,P,K kg soil^{-1}); G_i is partial soil loss in individual crop-stage periods (t ha^{-1}); $G_{r,i}$ is an average soil loss from parcel in *i*-year (t ha^{-1} year^{-1}); X_i is residual concentration of total N, P, K in soil in *i*-period (kg N,P,K ha^{-1}). The total concentrations of N, P, K in reservoir bottom sediments are computed as a weighted mean of the calculated average annual concentration of total N, P, K in transported soil particles from parcels in the watershed.

2.4. Statistical Methods

The validation of the proposed prediction model was carried out via a statistical assessment of the total N, P, and K concentrations in reservoir sediments calculated during the 10-year crop rotation system and determined nutrient concentrations in the Kl'usov small water reservoir.

It is natural to suppose that the random variables C_{Xy}—representing pertinent concentrations with X being N, P, or K, and y being either v for the values calculated by the proposed model or m for the measured data—exhibit a lognormal distribution. Thus the data should be logarithmically transformed.

In order to infer the proposed distribution a conformity test can be performed. The inference about the normality can be done using the Shapiro–Wilk (SW) test [44] for the transformed data. The hypothesis H_0: $C_{Xy} \sim \mathcal{N}(\mu(C_{Xy}), \sigma^2(C_{Xy}))$ is tested with respect to the alternative H_1: $C_{Xy} \sim$ non $\mathcal{N}(\mu(C_{Xy}), \sigma^2(C_{Xy}))$. The test statistic W for the used test is given as

$$W = \frac{\left(\sum\limits_{i=1}^{n} a_i x_i \right)^2}{\sum\limits_{i=1}^{n} (x_i - \bar{x})^2} \tag{4}$$

where the values in a pertinent dataset are denoted x_i the parameters of the test a_i are obtained from the table data for the SW test [45], which in the present case of $n = 20$ samples are

$$a_i = -a_{21-i}, i = 1, 2, \ldots 10, a_{11} = 0.0140, a_{12} = 0.0422, a_{13} = 0.0711, a_{14} = 0.1013, a_{15} = 0.1334, \tag{5}$$
$$a_{16} = 0.1686, a_{17} = 0.2085, a_{18} = 0.2565, a_{19} = 0.3211, a_{20} = 0.4734.$$

The critical value for the test statistic with a significance level 0.05 is $W_{crit} = 0.905$. If W is less than W_{crit}, the null hypothesis is rejected.

If, as assumed, the normality is not rejected at the 0.05 significance level, the random variables' means for each element can be compared—i.e., the calculated v and measured m—using the two-sample t-test (with the hypotheses H_0:$\mu v = \mu m$ vs. H_1:$\mu v \neq \mu m$), while for a comparison of the variances the two-sample F-test (with the hypotheses H_0:$\sigma v^2 = \sigma m^2$ vs. H_1:$\sigma v^2 \neq \sigma m^2$) is applied [46,47]. In all the cases, the significance level is set to the value 0.05. The parameters μ and σ are estimated by the sample mean $d\bar{x}$ and sample standard deviation S, respectively.

3. Results and Analysis

3.1. Field Measurements

Chemical analyses showed that measured N, P, and K concentrations in collected average soil samples from parcel 1004/1 ranged from 0.11 to 0.22% for total N, from 0.055 to 0.082% for total P, and from 1.72 to 1.91 for total K depending on collection period (different rates and date of fertilizer application, plant cover, crop uptake rates). Determined (measured) concentrations in soil samples from parcel 2001/1 ranged from 0.12 to 0.23% for total N, from 0.049 to 0.065% for total P, and from 1.67 to 1.85 for total K.

Measured N, P, and K concentrations in composite sediment samples are given in Table 1.

Table 1. Measured concentrations of total N, P, and K in reservoir sediments.

Sample	N (%)	P (%)	K (%)	Sample	N (%)	P (%)	K (%)
S1	0.260	0.112	2.420	S12	0.240	0.101	2.400
S2	0.240	0.113	1.980	S13	0.250	0.103	2.420
S3	0.230	0.066	1.960	S14	0.250	0.108	1.880
S4	0.220	0.066	1.690	S15	0.150	0.049	1.940

Table 1. *Cont.*

Sample	N (%)	P (%)	K (%)	Sample	N (%)	P (%)	K (%)
S5	0.220	0.067	2.050	S16	0.160	0.049	1.990
S6	0.200	0.090	2.030	S17	0.180	0.052	2.020
S7	0.170	0.049	1.710	S18	0.170	0.061	2.020
S8	0.160	0.070	2.200	S19	0.160	0.055	2.000
S10	0.230	0.086	2.320	S20	0.180	0.059	2.040
S11	0.230	0.103	2.370	S21	0.190	0.077	2.330

3.2. Model

The dissolved concentrations of total N and P in solution from soil samples were very low (at levels 0.22–0.43% of total N and 0.45–0.86% of total P) and therefore the dissolved forms of the following elements were neglected in this model.

The proposed model takes into account the calculation of total N, P, and K content adsorbed on detached and transported soil particles via water erosion.

The individual factors entering the equation are as follows. For the whole investigated territory (Bardejov district), it is considered with the constant value of factor R = 22.43 MJ ha^{-1} cm h^{-1} derived by Malisek [48]. K factor data is determined considering the soil texture in watershed and range from 0.25 to 0.40 t ha h ha^{-1} MJ^{-1} mm^{-1}. LS factor is determined for each outflow profile. The average $C_{r,i}$ values calculated are 0.25 for winter oilseed rape, 0.27 for triticale, 0.57 for corn silage, 0.17 for winter wheat, and 0.31 for spring barley. The P values are set at 1.0.

Based on the determination of a 'divided' C_i factor, partial soil loss G_i in individual crop-stage periods from studied parcels (1004/1 and 2001/1) are calculated and the results are given in Tables 2 and 3. Average annual soil loss ranged from 1.9 to 17.7 t ha^{-1} year^{-1} from 1004/1 parcel and from 5.34 to 27.7 t ha^{-1} year^{-1} for 2001/1 parcel according to used crop and management practices during the season. However, long term average annual soil loss is 8.3/14.3 t ha^{-1} year^{-1} from 1004/1; 2001/1 parcels depending on the crop and parcel length and slope gradient.

Table 2. Partial and average annual soil losses from 1004/1 parcel.

Year/Crop	G_i (t ha^{-1})					$G_{r,i}$ (t ha^{-1} Year^{-1})
	Crop-Stage Period *					
	1	2	3	4	5	
1/winter oilseed rape	3.74	1.88	0.14	1.22	0.78	7.75
2/triticale	5.41	0.36	0.08	1.55	0.75	8.16
3/corn silage	3.41	3.50	5.35	5.47	-	17.73
4/winter wheat	-	0.03	0.06	1.75	0.07	1.91
5/winter oilseed rape	1.96	1.34	0.11	1.26	0.77	5.45
6/winter wheat	5.24	0.19	0.06	1.46	0.75	7.71
7/potatoes	3.85	1.44	4.40	5.52	-	15.22
8/winter wheat	0.04	0.15	0.08	1.42	0.76	2.44
9/spring barley	4.17	0.96	2.39	0.85	0.21	8.58
10/winter oilseed rape	3.74	2.24	0.15	1.15	0.78	8.06

* 1—seedbed preparation; 2—establishment; 3—development; 4—maturing crop; 5—stubble field.

Table 3. Partial and average annual soil losses from 2001/1 parcel.

Year/Crop	G_i (t ha^{-1})					$G_{r,i}$ (t ha^{-1} Year^{-1})
	Crop-Stage Period					
	1	2	3	4	5	
1/corn silage	0.55	3.67	10.57	12.88	0.06	27.73
2/spring barley	0.10	0.20	1.85	2.91	0.28	5.34
3/triticale	5.88	3.99	0.21	3.07	0.14	13.28
4/winter oilseed rape	2.94	8.31	0.48	2.60	1.59	15.93
5/triticale	10.29	0.87	0.18	2.98	1.55	15.86
6/pea	8.24	1.70	4.43	1.87	1.57	17.81
7/winter wheat	8.74	0.61	0.14	4.39	0.10	13.99
8/spring barley	0.49	0.82	2.73	2.42	1.55	8.02
9/winter rye	7.57	0.67	0.15	2.88	1.85	13.12
10/winter oilseed rape	7.35	0.79	0.21	2.32	1.61	12.28

In terms of land management, the Tisovec River catchment falls under the authority of the Kľušov agricultural cooperative and all parcels within the catchment, with similar topographical and soil characteristics, are managed in a similar way. Consequently, calculations of total N, P, and K concentrations in eroded soil particles in adsorbed form are realized for the same (two) parcels of arable land situated next to the Kľušov reservoir.

Information about nutrient inputs is provided by the agricultural cooperative in the studied area. The fertilizers used in the watershed are primarily NPK fertilizers (8.5% or 15% N) and ammonium nitrate (27% N). Data about plant nutrient uptake are calculated according to the average yield in the Tisovec catchment's area and the mean plant nutrient uptake values provided by The Central Control and Testing Institute in Agriculture. For crops commonly grown in our conditions, partial (divided into five periods) and average plant nutrient uptake during the crop-stage periods was designed (Table 4).

Table 4. Proposed partial and average plant nutrient uptake during the crop-stage periods.

Crop	(kg ha^{-1})	Crop-Stage Period					$\sum_1^5$
		1	2	3	4	5	
spring barley	N	0.00	15.90	17.75	20.00	0.00	53.65
	P	0.00	2.05	2.50	3.70	0.00	8.25
	K	0.00	18.65	21.80	19.55	0.00	60.00
winter wheat, triticale, winter rye	N	0.00	0.00	19.10	71.35	0.00	90.45
	P	0.00	0.00	1.25	10.75	0.00	12.00
	K	0.00	0.00	9.55	67.50	0.00	77.05
winter oilseed rape	N	0.00	9.90	68.90	30.00	0.00	108.80
	P	0.00	1.40	15.90	5.00	0.00	22.30
	K	0.00	6.20	88.10	7.70	0.00	102.00
pea	N	0.00	18.30	17.70	26.30	0.00	62.30
	P	0.00	1.90	1.80	2.60	0.00	6.30
	K	0.00	8.50	8.20	12.10	0.00	28.80
potatoes	N	0.00	18.27	17.66	53.00	0.00	88.93
	P	0.00	2.61	2.52	7.58	0.00	12.71
	K	0.00	23.49	22.71	68.15	0.00	114.35
corn silage	N	0.00	16.35	16.89	50.18	0.00	83.42
	P	0.00	2.18	2.19	6.75	0.00	11.12
	K	0.00	13.62	13.67	42.22	0.00	69.51

The total N, P, and K content in transported soil particles from arable land to reservoir is calculated. The background concentrations of the modelled elements in topsoils from parcels 1004/1 and 2001/1 were 0.16/0.18% for N, 0.068/0.054% for P, and 1.81/1.80 for K. The values of *SER* ranged from 0.97 to 1.63 depending on an average soil loss from parcels in *i*-year.

Finally, the total nutrient contents in reservoir sediments C_{Xv} (Table 5) are then computed as a weighted mean of the calculated average annual total nutrient concentrations in transported soil particles from investigated parcels. The area of 1004/1 parcel is 15.63 ha and 2001/1 parcel area is 18.46 ha.

Table 5. Calculated concentrations of total N, P, and K in reservoir sediments.

Year	C_{Xv} in Reservoir (%)		
	N	P	K
1	0.188	0.067	1.956
2	0.225	0.078	2.307
3	0.188	0.066	1.944
4	0.225	0.082	2.387
5	0.203	0.072	2.131
6	0.193	0.068	2.032
7	0.189	0.065	1.954
8	0.236	0.083	2.469
9	0.199	0.069	2.071
10	0.202	0.070	2.096

3.3. Statistical Tests and Interpretation

Calculated concentrations of total N, P, and K in reservoir sediments (C_{Xv}) were compared with chemical analyses obtained from collected average sediment samples (C_{Xm}).

First, the SW test is used to surmise the normality of the distributions. The SW test statistic *W* (4) is calculated for each set of the logarithmic data. The results are summarized in Table 6, where also the comparison with critical value set at a 0.05 W_{crit} significance level is introduced. In each case, the null hypothesis of the SW test is not rejected so that the samples of transformed data can be considered to have a normal distribution.

Table 6. The values of test statistic *W* (*v* for calculated, m for measured data) in testing of H_0 and H_1 hypotheses using the SW test of conformity and their comparison with the critical value.

Element	SW Test					
	W_v		W_{crit}	W_m		W_{crit}
N	0.9255	>	0.905	0.9088	>	0.905
P	0.9346	>	0.905	0.9064	>	0.905
K	0.9145	>	0.905	0.9138	>	0.905

The results of the comparison of two samples for each element are shown in Table 7, the test statistics T are compared with critical percentages of the pertinent distributions F and t, respectively [44,46].

The results imply that for the case of nitrogen the hypothesis H_0 should be rejected, because the value of the test statistic is larger than the critical percentage. For phosphorus and potassium, however, the hypothesis H_0 cannot be rejected. This result influences the mode of the *t*-test for comparing the means: For N it is used with an assumption of unequal variances, while for P and K equal variances are assumed.

Table 7. The test statistics and the *F*-test and *t*-test critical percentages for a comparison of variances in N, P, and K concentrations.

Element	Variance			Mean		
	T		Percentage	T		Percentage
N	2.7161	>	2.5089	0.4956	<	2.0345
P	2.2073	<	2.5089	0.0863	<	2.0227
K	0.7173	<	2.5089	0.4078	<	2.0227

The results in Table 7 do not permit the rejection of the hypothesis H_0 for any of the concentrations considered. Thus, it can be supposed that the means of both results sets—i.e., measured and calculated according to the model (2)—N, P and K do not vary significantly in total concentrations. This result is also confirmed by finding the two-sided confidence intervals (CI) for distribution means μ.

The CIs obtained for a 0.95 confidence level both for transformed and untransformed random variables were calculated. Numerical data is supported by graphs in Figure 2 which show calculated and measured concentrations of all elements and which also include CIs of pertinent distribution means.

It can be observed that the CIs are overlapping for nitrogen and phosphorus even when CIv is inside CIm, while for potassium it is reversed. The explanation of this strict inclusion takes into account the errors caused by the sample collection and its determination. The reversed situation for potassium seems to be caused by the two extreme data obtained by the model.

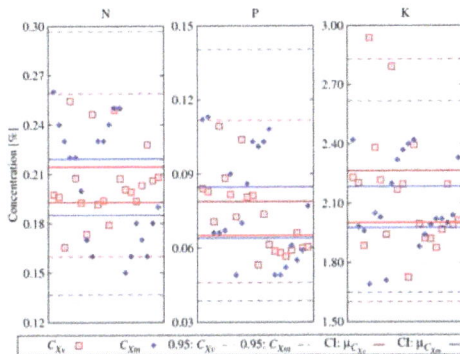

Figure 2. Graphical view of the statistical assessment of calculated (C_{Xv}) and measured (C_{Xm}) concentrations in sediments for X = N, P, or K.

4. Conclusions

This paper summarizes the results of a study aimed at the design of a mathematical model to predict the nitrogen, phosphorus, and potassium concentrations in reservoir bottom sediments. The proposed model has been developed and validated for the agricultural watershed of the Tisovec River in northeast of Slovakia, where the Kľušov reservoir is located. Based on the results described above, following conclusions can be summarized:

- Prediction model of sediment quality in the reservoir was specified depending on the main five periods of the cropping cycle and distribution of erosivity during a year.
- Total N, P, and K contents in bottom sediments were calculated considering soil loss using USLE equation supplemented with a determination of the average soil nutrient concentration in topsoils.

- The generalized plant cover factor value given in the USLE calculation was modified by its dividing into five crop-stage periods. To compute C, soil loss ratios were weighted according to annual distribution of erosivity.

- For selected crops (spring barley, winter wheat, triticale, winter rye, winter oilseed rape, corn silage, potatoes, and pea) an average plant nutrient uptake during five crop-stage periods was devised.

- Validation of the proposed model using t-test and F-test at a 0.05 significance level has shown that the suggested model can be used for predicting the content of total nitrogen, phosphorus, and potassium in reservoir sediments.

The study also reveals that the model may be considered a predictor for water management enterprises and agriculturists in the future.

Acknowledgments: This research has been supported by the Slovak Grant Agency for Science (Grant No. 1/0563/15) and the Slovak Cultural and Education Grant Agency (contract No. 073TUKE-4/2015).

Author Contributions: Natalia Junakova and Jozef Junak collected the samples; Natalia Junakova and Magdalena Balintova designed the model; Roman Vodicka performed validation of the designed model via a statistical assessment; Natalia Junakova wrote and edited the paper. All authors read and approved the final manuscript.

Conflicts of Interest: The authors declare no conflict of interest.

References

1. Boardman, J.; Poesen, J. (Eds.) Soil erosion in Europe: Major processes, causes and consequences. In *Soil Erosion in Europe*; Wiley: Chichester, UK, 2006; pp. 479–487, ISBN 978-047-085-9209.
2. Šestinova, O.; Findorakova, L.; Hančuľak, J.; Fedorova, E.; Špaldon, T. The Water Reservoir Ružín—Accumulation of Priority pollutants in Sediments in the Years 2010–2014. *Procedia Earth Planet. Sci.* **2015**, *15*, 844–848. [CrossRef]
3. Noor, H.; Fazli, S.; Alibakhshi, S.M. Evaluation of the relationships between runoff-rainfall-sediment related nutrient loss (A case study: Kojour Watershed, Iran). *Soil Water Res.* **2013**, *8*, 172–177.
4. Berc, J.; Bruce, J.; Easterling, D.; Groisman, P.Y.; Hatfield, J.; Hughey, B.; Johnson, G.; Kellogg, B.; Lawford, R.; Mearns, L.; et al. *Conservation Implications of Climate Change: Soil Erosion and Runoff from Cropland*; A Report from the Soil and Water Conservation Society: Ankeny, IA, USA, 2003.
5. European Environment Agency, Joint Research Centre. *The State of Soil in Europe—A Contribution of the JRC to the European Environment Agency's Environment*; State and Outlook Report—SOER 2010; Publications Office of the European Union: Luxembourg, 2012. [CrossRef]
6. European Environment Agency. *Climate Change, Impacts and Vulnerability in Europe 2012*; An Indicator-Based Report; European Environment Agency: Copenhagen, Denmark, 2012; ISBN 978-92-9213-346-7.
7. Eurostat. Agri-Environmental Indicator—Soil Erosion. Available online: http://ec.europa.eu/eurostat/statistics-explained/index.php?title=Agri-environmental_indicator_-_soil_erosion&oldid=263533#cite_note-1 (accessed on 20 November 2017).
8. Wisser, D.; Frolking, S.; Hagen, S.; Bierkens, M.F.P. Beyond peak reservoir storage? A global estimate of declining water storage capacity in large reservoirs. *Water Resour. Res.* **2013**, *49*, 5732–5739. [CrossRef]
9. World Commission on Dams. *Dams and Development: A New Framework for Decision Making*; Report of the World Commision on Dam; Earthscan Publications: London, UK, 2000.
10. Verstraeten, G.; Bazzoffi, P.; Lajczak, A.; Rādoane, M.; Rey, F.; Poesen, J.; Vente, J. Reservoir and Pond Sedimentation in Europe. In *Soil Erosion in Europe*; Boardman, J., Poesen, J., Eds.; John Wiley and Sons Ltd.: Chichester, UK, 2006; pp. 757–774, ISBN 978-0-470-85910-0.
11. Batuca, D.G.; Jordaan, J.M. *Silting and Desilting of Reservoirs*; CRC Press: Rotterdam, The Netherlands, 2000; ISBN 90-5410-477-5.
12. Syvitski, J.P.M.; Milliman, J.D. Geology, geography, and humans battle for dominance over the delivery of fluvial sediment to the coastal ocean. *J. Geol.* **2007**, *115*, 1–19. [CrossRef]
13. Koc, C. A study on sediment accumulation and environmental pollution of Fethiye Gulf in Turkey. *Clean Technol. Environ. Policy* **2012**, *14*, 97–106. [CrossRef]

14. Walling, D.E.; Webb, B.W. Erosion and Sediment Yield: A global overview. In *Erosion and Sediment Yield: Global and Regional Perspectives*; Walling, D.E., Webb, B.W., Eds.; IAHS Publications: Wallingford, UK, 1996; pp. 3–20, ISBN 978-0947571894.

15. Li, X.; Wang, B.; Yang, T.; Zhu, D.; Nie, Z.; Xu, J. Identification of soil P fractions that are associated with P loss from surface runoff under various cropping systems and fertilizer rates on sloped farmland. *PLoS ONE* **2017**, *12*, e0179275. [CrossRef] [PubMed]

16. Niu, X.Y.; Wang, Y.H.; Yang, H.; Zheng, J.W.; Zou, J.; Xu, M.N.; Wu, S.S.; Xie, B. Effect of Land Use on Soil Erosion and Nutrients in Dianchi Lake Watershed, China. *Pedosphere* **2015**, *25*, 103–111. [CrossRef]

17. Liu, X.; Li, Z.; Li, P.; Zhu, B.; Long, F.; Cheng, Y.; Wang, T.; Lu, K. Changes in carbon and nitrogen with particle size in bottom sediments in the Dan River, China. *Quat. Int.* **2015**, *380–381*, 305–313. [CrossRef]

18. Qian, J.; Zhang, L.P.; Wang, W.Y.; Liu, Q. Effects of Vegetation Cover and Slope Length on Nitrogen and Phosphorus Loss from a Sloping Land under Simulated Rainfall. *Pol. J. Environ. Stud.* **2014**, *23*, 835–843.

19. Zeng, S.C.; Su, Z.Y.; Chen, B.G.; Wu, Q.T.; Ouyang, Y. Nitrogen and phosphorus runoff losses from orchard soils in South China as affected by fertilization depths and rates. *Pedosphere* **2008**, *18*, 45–53. [CrossRef]

20. Young, R.A.; Onstad, C.A.; Bosch, D.D.; Anderson, W.P. Agricultural non-point source pollution model for evaluating agricultural watersheds. *J. Soil Water Conserv.* **1989**, *44*, 168–173.

21. Pimentel, D.; Burgess, M. Soil Erosion Threatens Food Production. *Agriculture* **2013**, *3*, 443–463. [CrossRef]

22. Ignatieva, N.V. Nutrient exchange across the sediment-water interface in the eastern Gulf of Finland. *Boreal Environ. Res.* **1999**, *4*, 295–305.

23. Estigoni, M.V.; Miranda, R.B.; Mauad, F.F. Hydropower reservoir sediment and water quality assessment. *Manag. Environ. Qual. Int. J.* **2017**, *28*, 43–56. [CrossRef]

24. Teodoru, C.; Wehrli, B. Retention of Sediments and Nutrients in the Iron Gate I Reservoir on the Danube River. *Biogeochemistry* **2005**, *76*, 539–565. [CrossRef]

25. Ongley, M. Sediment measurements. In *Water Quality Monitoring: A Practical Guide to the Design and Implementation of Freshwater Quality Studies and Monitoring Programmes*; Bartram, J., Balance, R., Eds.; UNEP/WHO: London, UK, 1996; Chapter 13, pp. 1–15, ISBN 0-419-22320-7.

26. Zeleňáková, M.; Čarnogurská, M.; Šlezingr, M.; Słys, D.; Purcz, P. A model based on dimensional analysis for prediction of nitrogen and phosphorus concentrations at the river station Ižkovce, Slovakia. *Hydrol. Earth Syst. Sci.* **2013**, *17*, 201–209. [CrossRef]

27. Ganasri, B.P.; Ramesh, H. Assessment of soil erosion by RUSLE model using remote sensing and GIS—A case study of Nethravathi Basin. *Geosci. Front.* **2016**, *7*, 953–961. [CrossRef]

28. Saha, S.K. Water and wind induced soil erosion assessment and monitoring using remote sensing and GIS. In *Satellite Remote Sensing and GIS Applications in Agricultural Meteorology*; Sivakumar, M.V.K., Roy, P.S., Harmsen, K., Saha, S.K., Eds.; World Meteorological Organization: Geneva, Switzerland, 2004; pp. 315–330.

29. Baudron, F.; Thierfelder, C.; Nyagumbo, I.; Gérard, B. Where to Target Conservation Agriculture for African Smallholders? How to Overcome Challenges Associated with its Implementation? Experience from Eastern and Southern Africa. *Environments* **2015**, *2*, 338–357. [CrossRef]

30. Diodato, N.; Guerriero, L.; Bellocchi, G. Modeling and Upscaling Plot-Scale Soil Erosion under Mediterranean Climate Variability. *Environments* **2017**, *4*, 58. [CrossRef]

31. Ma, L.; Bu, Z.H.; Wu, Y.H.; Kerr, P.G.; Garre, S.; Xia, L.Z.; Yang, L.Z. An integrated quantitative method to simultaneously monitor soil erosion and non-point source pollution in an intensive agricultural area. *Pedosphere* **2014**, *24*, 674–682. [CrossRef]

32. Bosco, C.; de Rigo, D.; Dewitte, O.; Poesen, J.; Panagos, P. Modelling soil erosion at European scale: Towards harmonization and reproducibility. *Nat. Hazards Earth Syst. Sci.* **2015**, *15*, 225–245. [CrossRef]

33. Huang, T.C.C.; Lo, K.F.A. Effects of Land Use Change on Sediment and Water Yields in Yang Ming Shan National Park, Taiwan. *Environments* **2015**, *2*, 32–42. [CrossRef]

34. Mokhtar, N.H.; Gofar, N.; Kassim, A. *Combining Design Methodologies for the Development of a Practical and Effective Approach to Erosion Control Systems*; Project Report Vot 74179; Universiti Teknologi Malaysia: Johor Bahru, Malaysia, 2006.

35. National Research Council (Committee on Long-Range Soil and Water Conservation, Board on Agriculture). *Soil and Water Quality: An Agenda for Agriculture*; National Academy Press: Washington, DC, USA, 1993; p. 542, ISBN 978-0-309-04933-7.

36. Junakova, N.; Junak, J. Sustainable Use of Reservoir Sediment through Partial Application in Building Material. *Sustainability* **2017**, *9*, 852. [CrossRef]

37. Michalec, B. Qualitative and quantitative assessment of sediments pollution with heavy metals of small water reservoirs. In *Soil Health and Land Use Management*; Hernandez-Soriano, M.C., Ed.; InTech: Rijeka, Croatia, 2012; pp. 255–278, ISBN 978-953-307-614-0.

38. Slovak Water Management Enterprise. *Operational Manual of the Klusov Water Reservoir (in Slovak)*; Slovak Water Management Enterprise: Trebišov, Slovak Republic, 2005.

39. Mahler, R.L.; Tindall, T.A. *Soil sampling. Bulletin 704 (Revised)*; College of Agriculture, University of Idaho: Moscow, ID, USA; U.S. Department of Agriculture: Washington, DC, USA, 1994.

40. Ministry of the Environment of the Slovak Republic. *Methodological Instruction No. 549/98-2 for Risk Assessment Posed by Contaminated Sediments in Streams and Water Reservoirs (in Slovak)*; Ministry of the Environment of the Slovak Republic: Bratislava, Slovak Republic, 1998.

41. Wischmeier, W.H.; Smith, D.D. Predicting Rainfall Erosion Losses—A Guide to Conservation Planning. In *Agriculture Handbook No. 537*; US Department of Agriculture: Beltsville, MD, USA, 1978.

42. Karaburun, A. Estimation of C factor for soil erosion modeling using NDVI in Buyukcekmece watershed. *Ozean J. Appl. Sci.* **2010**, *3*, 77–85.

43. Menzel, R.G. Enrichment ratios for water quality modeling. In *CREAMS, A Field Scale Model for Chemicals, Runoff, and Erosion from Agricultural Management Systems*; Knisel, W., Ed.; US Department of Agriculture: Phoenix, AZ, USA, 1980; Volume 3: Supporting Documentation, Conservation Report No. 26; pp. 486–492.

44. Shapiro, S.S.; Wilk, M.B. An analysis of variance test for normality (complete samples). *Biometrika* **1965**, *52*, 591–611. [CrossRef]

45. Zaiontz, C. Real Statistics Using Excel. Available online: http://www.real-statistics.com/tests-normality-and-symmetry/statistical-tests-normality-symmetry/shapiro-wilk-test/ (accessed on 14 December 2016).

46. Chakravarti, I.M.; Laha, R.G.; Roy, J. *Handbook of Methods of Applied Statistics*; John Wiley and Sons: New York, NY, USA, 1967; ISBN 978-1135545802.

47. Bhattacharya, G.K.; Johnson, R.A. *Statistical Concepts and Methods*; John Wiley and Sons: New York, NY, USA, 1977; ISBN 9780471072041.

48. Malisek, A. Evaluation of the rainfall and runoff factor by geographic location. *Geogr. Čas.* **1990**, *42*, 410–422.

environments

MDPI

Article

Greenhouse Gas Emission Assessment from Electricity Production in the Czech Republic

Simona Jursová [1],*, Dorota Burchart-Korol [2], Pavlína Pustějovská [1], Jerzy Korol [3] and Agata Blaut [3]

[1]   Centre Enet, VSB—Technical University of Ostrava, 708 33 Ostrava, Czech Republic;
     pavlina.pustejovska@vsb.cz
[2]   Faculty of Transport, Silesian University of Technology, 40-019 Katowice, Poland;
     dorota.burchart-korol@polsl.pl
[3]   Department of Material Engineering, Department of Water Protection, Central Mining Institute,
     40-166 Katowice, Poland; jkorol@gig.eu (J.K.); ablaut@gig.eu (A.B.)
*   Correspondence: simona.jursova@vsb.cz; Tel.: +420-597-325-421

Received: 24 November 2017; Accepted: 22 January 2018; Published: 23 January 2018

Abstract: The paper deals with the computational life cycle assessment (LCA) model of electricity generation in the Czech Republic. The goal of the paper was to determine the environmental assessment of electricity generation. Taking into account the trend of electricity generation from 2000 to 2050, the paper was focused on electricity generation evaluation in this country in view of its current state and future perspectives. The computational LCA model was done using the Intergovernmental Panel on Climate Change (IPCC) method, which allowed the assessment of greenhouse gas emissions. For the assessment, 1 Mega-watt hour of the obtained electricity (MWhe) was used as a functional unit. The cradle-to-gate approach was employed. The system boundary covered all the technologies included in the electricity mix of the country. Resulting from the analysis, the solids, lignite in particular, was assessed as an energy source with the most negative impact on the emissions of greenhouse gas. This article results from international cooperation of a Czech-Polish team in the field of computational LCA models. It presents partial results of the team cooperation which serves as a base for following comparison of Czech and Polish systems of electricity generation.

Keywords: environment; electricity; LCA assessment; Czech Republic

1. Introduction

Searching for cost effective and environmentally friendly energy sources has become more critical with increasing concerns about sustainability [1]. Electricity is one of the main contributors to global environmental impacts. It is one of the main sources of environmental burden in several sectors such as buildings or information and communication technology [2]. Electricity supply is often highlighted as a significant hot spot in Life cycle assessment (LCA) results for a majority of product and service life cycles [3]. The method of LCA is one of the most effective tools being used to provide a material and energy balance over the entire life of a material, product, technology, or service, determining its interaction with its environment, and assessing its impacts on the environment [4]. It has its roots in a number of studies conducted in the 1960s and 70s that aimed at optimizing energy consumption in a context where the latter represented a restraint for the industry. Since then, LCA has been further developed and standardized but its core aim has always remained to understand the overall environmental impacts of a product or system along its full lifecycle (from the extraction of the necessary primary resources, to end-of-life disposal, and recycling where applicable) [5]. There are studies, following LCA analysis, which are aimed at possibilities of further product utilization, e.g., into segments of construction materials [6–8].

The life cycle assessment (LCA) can support the elaboration of policies to meet global or regional challenges. It can also allow for the identification of hotspots and the refining of existing energy policies, e.g., supporting amendments in national emission standards and prioritizing or targeting specific energy sources and technologies identified as important causes of environmental impact in the countries considered. The LCA is a systemic tool and is, thus, highly relevant for evaluating long-term electricity trajectories, which can encompass all electricity supply systems and their interactions with other systems and society at large [9,10].

In the field of power engineering, this method has been used globally. The reference literature shows results of LCA of individual energy sources and technologies. The assessment of electricity system generation in general view of energy sources has been carried out for many countries such as China, UK, USA, Portugal, Romania, Poland, Mexico, Mauritius, and Tanzania [11–17]. Some authors assessed electricity generation in view of greenhouse gas emissions (GHGs) [18,19]. Some authors studied on environmental impact of electricity generation from renewable resources [20–27]. There are also papers that have assessed clean and innovative power technologies [28–31]. In the case of the Czech Republic, there is a lack of papers concerning the LCA of energy systems. Only the results of the LCA analyses of municipal waste management in the Czech Republic are presented in literature [32]. This paper's objective is to provide an environmental assessment of energy sources used for electricity generation in the Czech Republic. Its purpose is the assessment of applied energy sources and interpretation of their environmental effect in current and future perspectives.

2. Materials and Methods

Environmental assessment has been carried out for electricity generation in the Czech Republic; a country located in central Europe with a population of over 10 million inhabitants. The country's surface area is 78,866 km^2. The Czech Republic is one of the most industrialized economies and developed in the Central and Eastern Europe. The Czech electricity market is characterized by a very positive attitude towards nuclear power, a dominant position of lignite in the Czech Republic electricity generation, and a strong role of electricity export since the Czech Republic ranks sixth in the world and fourth in Europe in electricity exports [33]. The dominant source of energy in the Czech Republic is solids (hard coal and lignite), which constitutes 39% of the total primary energy supply. The computational LCA model was based on International Organization for Standardization (ISO) ISO 14,040:2006 using SimaPro 8 software (Pré Sustainability, Amersfoort, The Netherlands) with Ecoinvent 3 database. The authors changed electricity mix from ecoinvent according to data from [34]. Reference [35] describes the principles and framework for LCA, including four stages which are presented on the Figure 1.

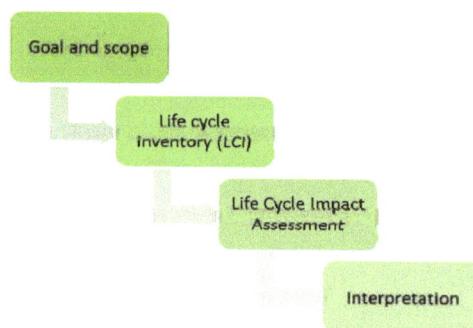

Figure 1. Stages of life cycle assessment based on [35].

The cradle-to-gate approach was employed. The functional unit was applied as the production of 1 MWh of electricity. All input and output are indicated based on this functional unit. Environmental assessment was made using the LCA method, based on the IPCC method, allowing the presentation of greenhouse gas emissions (GHGs) with life cycle approach. IPCC method was developed by the Intergovernmental Panel on Climate Change (IPCC). The IPCC publishes Global Warming Potentials (GWPs) [36]. GHG emissions from analyzed systems were quantified per unit of electricity. The impact category refers to GHGs and expresses the radiative forcing of greenhouse gas emissions over a 100-year horizon, expressed in kilograms of CO_2 equivalent. The life cycle greenhouse gas emission involves analysis the GWPs of energy sources through life cycle assessment of energy sources for electricity generation.

For the purpose of computational environmental life cycle assessment model, data for the current and future electricity generation in the Czech Republic were identified in European and global energy bases. Figure 2 shows the inventory data of electricity production from 2000 to 2050 that was required to perform the computational model.

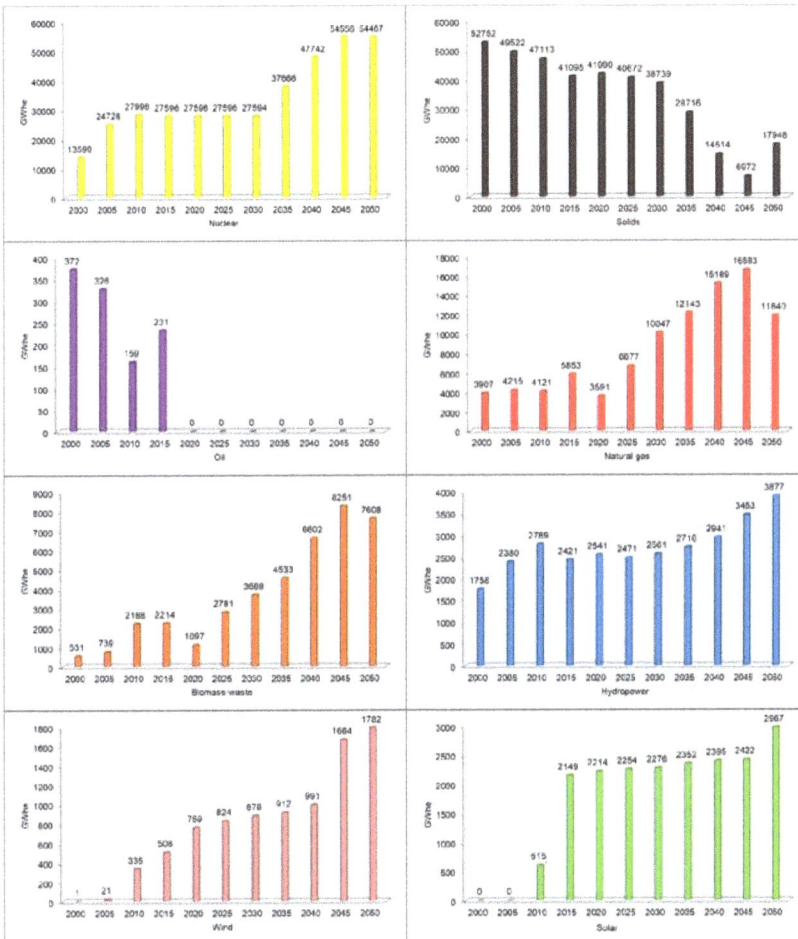

Figure 2. Live cycle inventory data for analysis, based on [34,37].

In Czech Republic, the main energy sources for electricity are solids (lignite and hard coal) and nuclear energy. The trend in 2000–2050 shows a drop from 52,752 Giga-watt hour of the obtained electricity (GWhe) to 17,948 GWhe in generation from coal and an increase from 13,590 GWhe to 54,467 GWhe in nuclear energy. Coal is the only significant indigenous energy resource in the Czech Republic. The country's proven coal reserves have been estimated at some 880 million tonnes. Hard coal accounts for 10%, while lignite presents more than 90% of these reserves. Lignite provides an important contribution to the country's energy supply. The largest lignite deposit in the Czech Republic holds reserves of 750 million tonnes of good quality coal with an energy content of up to 17,500 kJ/kg. These reserves are estimated to last for the next 100 years, subject to mining limits set in 1991 [38].

The nuclear sources in the Czech Republic are primarily used for generation of electricity. These now supply 32% of all electricity produced in the country. There are two nuclear power plants in the Czech Republic, in Dukovany and in Temelin. One of the priorities of the State Energy Policy is the completion of the construction of additional nuclear power units to produce around 20 Terra-watt hour (TWh by 2035, extending the lifetime of the existing four units in the Dukovany power plant and later the possible construction of another unit to compensate for the commissioning of this plant. In the long term strategy, nuclear energy is supposed to provide electricity in excess of 50% to replace a large proportion of the coal sources. Sites for future additional nuclear power stations in the Czech Republic after 2040 are being explored and prepared [39].

Oil and gas deposits are insignificant in the Czech Republic. From 2020, oil is not expected to be used in the Czech Republic for the production of electricity. The current ratio of gas to electricity generation is roughly 3%. The domestic energy system is practically entirely dependent on imports of this energy commodity. The dominant supplier is the Russian Federation, together with Norway. In long term strategy, electricity from natural gas is expected to be generated in cogeneration. The Czech Republic in its energy policy assumes the development of so-called micro cogeneration sources, primarily using natural gas [39]. The total proportion of gas in the energy mix therefore rises after 2020. In 2050, gas fired generation is expected to be partially replaced by solids and decline to 11,840 GWhe.

The decline in the ratio of domestic energy sources to the electricity generation from primary energy sources inevitably leads to the development of low-carbon ones. The Action Plan for Biomass in the Czech Republic legislatively supports the development of electricity generation from biomass and waste until domestic potential will have been exhausted [40]. In long term strategy, the biodegradable municipal waste is expected to replace primary sources. Currently, 500 kg of municipal waste is generated per inhabitant of the Czech Republic per year. The amount of waste rises as the population's purchasing power increases. Nowadays, there are only three waste energy utilization facilities in the Czech Republic, with a processing capacity of 654 tonnes per year [39].

Czech geographic conditions allow the installation of renewable energy plants which make use of weather, like wind or sun. However, due to the natural environment, these types of plants yield only average results. Although renewable energy sources are not competitive on their own and the support for every MWh of the green energy produced and supplied to the grid is necessary, the increasing trend in electricity generated in wind and solar plants is visible in following years until 2050. In 2015, there was a sharp rise in the use of solar energy for electricity production due to the disproportionate amount of support for newly built commercial plants from 2006–2010. Before legislation reacted to the photovoltaic boom (by the end of 2010), the Czech installed solar capacity rose from 40 MW in 2008 to 1960 MW in 2010. The Czech subsidy for solar electricity dropped from an initial 15,565 Czech crowns (CZK)/MWh in 2006 to zero for newly built commercial photovoltaic plants in 2014 [33].

Distribution of energy sources on electricity generation in the Czech Republic is in Figure 3. Renewable energy represents 12% of the total amount of 82.6 TWh of electricity generated in 2015. In comparison with the average of International Energy Agency countries, it is a half share. In the European Union (EU) in 2015, the main sources of electricity included nuclear energy 867,402 GWhe (which constituted 26.68%), solids 846,834 GWhe (which constituted 26.04%), and natural gas

566,075 GWhe (which constituted 17.41%). Renewable energy sources accounted for 28.2% of electricity generation in total.

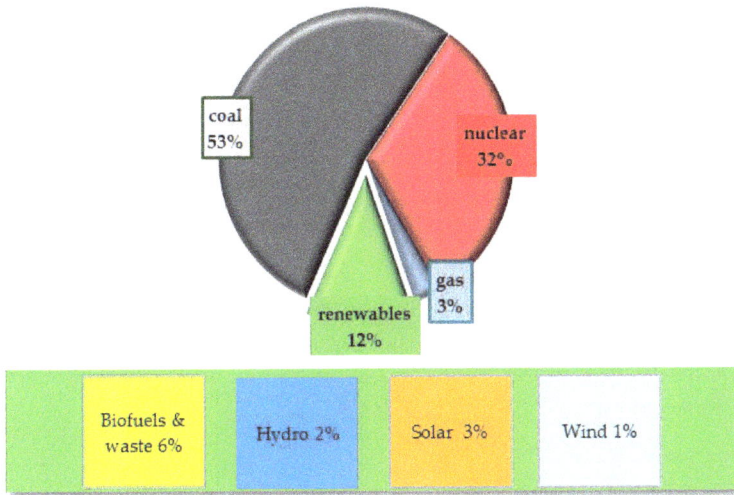

Figure 3. Distribution of energy sources on electricity generation in Czech Republic based on [41].

In the EU, wind and hydro plants provide the largest contribution from these sources, supplying 19% of gross electricity generation in 2015. Generation from solar sources contributes 3%. The investments into solar plants are expected in Southern Europe to reach an installed capacity of 183 GW by 2030 and 299 GW by 2050 [34]. Also, the use of biomass and waste combustion for power generation is expected to increase over time, both in pure biomass plants and cofiring applications in solid fuel plants. The current biomass and waste contribution to electricity generation in European Union is 6%. The share of geothermal electricity generation in long time European strategy remains at 0.2%. In the European Union, gas plays a crucial role in the context of emission reduction targets and increased implementation of variable renewable energy sources. Gas-fired generation is a back-up technology for variable energy sources.

Electricity generation from solids in European Union declines significantly in the projection period from 2000–2050. The same trend is recorded in energy strategy of the Czech Republic. At the end of projection time, share of solids in gross electricity production increases in European Union from 5.9% in 2045 to 6.1% in 2050; in the Czech Republic from 7.4% even to 17.9% [37]. It is explained by economically driven investments into clean coal technologies and technology of carbon capture and storage (CCS) which has taken place in the long term in countries such as Czech Republic with substantial solid generation and endogenous resources. According to the EU Reference Scenario, by 2050 in the European Union, more than half of solid-fuelled generation will be produced from facilities with installed CCS technologies.

3. Results

The comparative analysis of life cycle greenhouse gas emissions of electricity production in the Czech Republic is presented in Figure 4. The GHGs for the electricity generation was 917 kg CO_2 eq/MWh in 2000, while in 2050 the potential impact on GHG will be 331 kg CO_2 eq/MWh. The value of GHG emissions from 2000 over the past 50 years decreased by 66%. Table 1 then sums up calculated GHG emissions from individual energy sources.

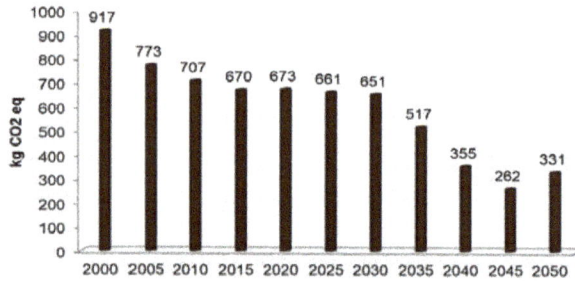

Figure 4. Analysis of greenhouse gas emissions of electricity production in the Czech Republic based on Intergovernmental Panel on Climate Change (IPCC) method.

Table 1. Greenhouse gas emission electricity production from individual energy sources in the Czech Republic in kg CO_2 eq.

Year	Total	Nuclear	Hard Coal	Lignite	Oil	Natural Gas	Biomass Waste	Hydro	Wind	Solar
2000	917	1.45	112.69	746.98	6.15	48.97	0.43	0.09	0.00	0.00
2005	773	2.35	94.15	624.03	4.79	47.01	0.53	0.11	0.00	0.00
2010	707	2.56	86.02	570.10	2.24	44.14	1.50	0.12	0.04	0.62
2015	670	2.62	77.99	516.98	3.39	65.17	1.58	0.11	0.07	2.25
2020	673	2.69	81.97	543.32	0.00	41.13	0.80	0.12	0.11	2.39
2025	661	2.58	76.08	504.22	0.00	73.27	1.95	0.11	0.11	2.33
2030	651	2.51	70.36	466.33	0.00	107.05	2.50	0.11	0.12	2.28
2035	517	3.30	50.23	332.96	0.00	124.62	2.98	0.11	0.12	2.27
2040	355	4.12	25.02	165.80	0.00	153.58	4.27	0.12	0.12	2.28
2045	262	4.53	11.57	76.65	0.00	161.37	5.14	0.14	0.20	2.22
2050	331	4.22	27.81	184.39	0.00	107.67	4.43	0.14	0.20	2.54

An analysis of the greenhouse gas emissions for each source of electricity in the Czech Republic was done in order to determine the environmental determinants of the impacts. The determinants of the GHG emissions of current and future electricity generation in the Czech Republic are presented in Figure 5.

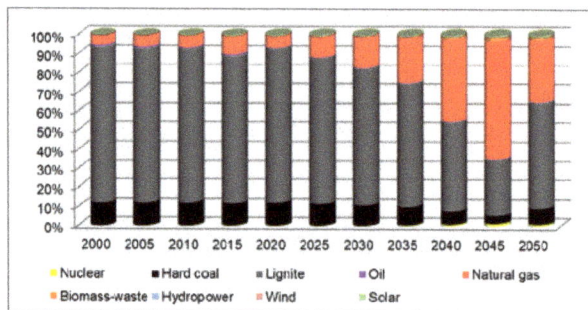

Figure 5. Determinants of greenhouse gas (GHG) emissions of current and future electricity generation in the Czech Republic based on IPCC method.

4. Discussion

The environmental assessment of Czech electricity sources shows that the greenhouse gas (GHG) emissions are determined by generation from solids, in particular lignite. The share of solids in the electricity generation system in the Czech Republic is mostly lignite over the analyzed period. It shows that electricity generation from lignite is associated with high impact in greenhouse gas emissions. The analyses that were carried out have shown that, despite the decreasing tendency of environmental impact, a slight increase related to the emissions of greenhouse gas was indicated between 2045 and 2050. This is due to an increase in the share of lignite and hard coal in the electricity generation structure in 2050 compared to 2045, as shown in Figure 2. It is predicted to increase from 67,972 GWhe to 17,948 GWhe. The increasing tendency is expected in relation with implementation of clean coal technologies such as coal gasification in integrated gas combined cycle (IGCC), in atmospheric fluidized bed (AFB), or pressurized fluidized-bed combustion (PFBC). These technologies are regarded to be more environmentally friendly in comparison to conventional coal power plants due to their high thermal efficiency, capability to process low grade coal, and low non-carbon GHG emissions. In this period, the technology of carbon capture storage (CCS) is supposed to be implemented in some power plants to prevent CO_2 from entering the atmosphere. The vision of establishment of efficient high-tech technologies changes the approach to solids, resulting in increasing electricity generation from this indigenous resource in the Czech Republic. Electricity generation from solids after 2050 remain a leading source in GHG emissions predicted to emit 212.2 $kgCO_{2eq}$ (Table 1) for 17,948 GWhe (Figure 2) of electricity in 2050. Since 2000, GHG emissions from solids have decreased four times. The amount of predicted electricity from the source is expected to decrease just three times in this period.

This study has shown that the share of nuclear energy in electricity generation will increase in the Czech Republic. Despite the increase of nuclear energy, it is not associated with an increase in greenhouse gas emissions. In 2000, 13,590 GWhe of electricity was produced in Czech nuclear plants and only 1.45 $kgCO_{2eq}$ was emitted. The nuclear electricity presents about 32% in electricity mix of the Czech Republic. GHG emissions 2.62 $kgCO_{2eq}$ indicated for this distribution are insignificant. By 2050, GHG emissions from nuclear source are predicted to be 4.22 $kgCO_{2eq}$ for a quadruple increase in electricity generated at 54,467 GWhe. Electricity generation from solids reached a similar value (52,752 GWhe) in 2000 and presented 859.67 $kgCO_{2eq}$ of GHG emissions.

The trend similar to electricity from nuclear plants is recorded in the case of hydropower energy, solar, and wind. Electricity generation from these sources has an increasing tendency which does not result in a significant increase in GHG emissions. In consideration of renewable resources, the least ecological one is solar energy. An analysis of environmental determinants has shown that electricity generation in solar plants relates to largest GHG emissions in group of green energy sources. GHG emissions from solar energy are expected 2.54 $kgCO_{2eq}$ for 2967 GWhe of electricity in 2050. To compare, similar amount of GHG emissions from nuclear energy 2.56 $kgCO_{2eq}$ was in 2010 for 27,998 GWhe of electricity.

It has been reported that an increase in the natural gas source for the production of electricity affects the increase of GHG emissions. The ratio in GHG emissions of natural gas is expected to increase significantly from 2040 to 2050. In 2045, emissions are expected to reach 161.37 $kgCO_{2eq}$. In this year, the electricity generation from natural gas is predicted to overcome emissions from solid sources related to the significant decline in electricity production from solids predicted for this year.

5. Conclusions

Based on analysis, inventory data for electricity generation in the Czech Republic were identified. The main energy sources in the Czech Republic are nuclear energy, lignite, and hard coal. The trend in the years 2000–2050 shows an increase in nuclear energy and a decrease in electricity generation from coal. Oil is not expected to be used in the Czech Republic for the production of electricity from 2020. An increase in generation from wind, gas, solar, and biomass is expected in the following years until 2050.

Despite the reduction in the share of lignite and hard coal in the electricity generation system in the Czech Republic, the share of solids has the greatest impact on the greenhouse gas emissions.

The obtained results show that the determinant of the greenhouse gas emissions from electricity generation systems is electricity from lignite. Despite a significant increase of nuclear power in electricity generation systems, this source does not affect the greenhouse gas emissions.

Acknowledgments: This paper was written within the framework of the project LO1404: Sustainable development of Centre of Energy Units for Utilization of non-Traditional Energy Sources (ENET) and supported by project Interreg V-A Czech Republic–Poland, Microprojects Fund 2014-2020 in the Euroregion Silesia reg. no. CZ.11.4.120/0.0/0.0/16_013/0000653.

Author Contributions: Simona Jursová and Dorota Burchart-Korol conceived the idea and designed the framework for the analysis. Agata Blaut and Pavlína Pustějovská prepared inventory data analysis. Dorota Burchart-Korol performed software assessment; Agata Blaut contributed analysis tools. Jerzy Korol and Pavlína Pustějovská interpreted the results. Simona Jursova wrote the paper. Dorota Burchart-Korol and Jerzy Korol summed up the conclusions.

Conflicts of Interest: The authors declare no conflict of interest.

References

1. Weldu, Y.W.; Assefa, G. The search for most cost-effective way of achieving environmental sustainability status in electricity generation: Environmental life cycle cost analysis of energy scenarios. *J. Clean. Prod.* **2017**, *142*, 2296–2304. [CrossRef]
2. Astudillo, M.F. Life cycle inventories of electricity supply through the lens of data quality: Exploring challenges and opportunities. *Int. J. Life Cycle Assess.* **2017**, *22*, 374–386. [CrossRef]
3. Treyer, K.; Bauer, C. Life cycle inventories of electricity generation and power supply in version 3 of the ecoinvent database—Part II: Electricity markets. *Int. J. Life Cycle Assess.* **2014**. [CrossRef]
4. Asif, M.; Dehwah, A.H.A.; Ashraf, F.; Khan, H.S.; Shaukat, M.M.; Hassan, M.T. Life Cycle Assessment of a Three-Bedroom House in Saudi Arabia. *Environments* **2017**, *4*, 52. [CrossRef]
5. Raugei, M.; Leccisi, E. A comprehensive assessment of the energy performance of the full range of electricity generation technologies deployed in the United Kingdom. *Energy Policy* **2016**, *9*, 46–59. [CrossRef]
6. Václavík, V.; Valíček, J.; Dvorský, T.; Hryniewicz, T.; Rokosz, K.; Harničárová, M.; Kušnerová, M.; Daxner, J.; Bendová, M. A method of utilization of polyurethane after the end of its life cycle. *Rocznik Ochrony Środowiska* **2012**, *14*, 96–106.
7. Dvorský, T.; Václavík, V.; Šimíček, V.; Břenek, A. Research of the Use of Waste Rigid Polyurethane Foam in the Segment of Lightweight Concretes. *Inzynieria Mineralna* **2015**, *36*, 51–56.
8. Břenek, A.; Öchsner, A.; Václavík, V.; Altenbach, H.; Dvorský, T.; Daxner, J.; Dirner, V.; Bendová, M.; Harničárová, M.; Valíček, J. Capillary Active Insulations Based on Waste Calcium Silicates. *Adv. Struct. Mater.* **2015**, *70*, 177–188.
9. Günkaya, Z.; Özdemir, A.; Özkan, A.; Banar, M. Environmental Performance of Electricity Generation Based on Resources: A Life Cycle Assessment Case Study in Turkey. *Sustainability* **2016**, *8*, 1097. [CrossRef]
10. Laurent, A.; Espinosa, N. Environmental impacts of electricity generation at global, regional and national scales in 1980–2011: What can we learn for future energy planning? *Energy Environ. Sci.* **2015**, *8*, 689–701. [CrossRef]
11. Ou, X.; Yan, X.; Zhang, X. Life-cycle energy consumption and greenhouse gas emissions for electricity generation and supply in China. *Appl. Energy* **2011**, *88*, 289–297. [CrossRef]
12. Garcia, R.; Marques, P.; Freire, F. Life-cycle assessment of electricity in Portugal. *Appl. Energy* **2014**, *134*, 563–572. [CrossRef]
13. Stamford, L.; Azapagic, A. Life cycle sustainability assessment of UK electricity scenarios to 2070. *Energy Sustain. Dev.* **2014**, *23*, 194–211. [CrossRef]
14. Peiu, N. Life cycle inventory study of the electrical energy production in Romania. *Int. J. Life Cycle Assess.* **2007**, *12*, 225–229. [CrossRef]
15. Felix, M.; Gheewala, S.H. Environmental assessment of electricity production in Tanzania. *Energy Sustain. Dev.* **2012**, *16*, 439–447. [CrossRef]
16. Santoyo-Castelazo, E.; Gujba, H.; Azapagic, A. Life cycle assessment of electricity generation in Mexico. *Energy* **2011**, *36*, 1488–1499. [CrossRef]
17. Brizmohun, R.; Ramjeawon, T.; Azapagic, A. Life cycle assessment of electricity generation in Mauritius. *J. Clean. Prod.* **2014**, *16*, 1727–1734. [CrossRef]

18. Jones, C.; Gilbert, P.; Raugei, M.; Mander, S.; Leccisi, E. An approach to prospective consequential life cycle assessment and net energy analysis of distributed electricity generation. *Energy Policy* **2017**, *100*, 350–358. [CrossRef]

19. Hondo, H. Life cycle GHG emission analysis of power generation systems: Japanese case. *Energy* **2005**, *30*, 2042–2056. [CrossRef]

20. Kannan, R.; Leong, K.C.; Osman, R.; Ho, H.K. Life cycle energy, emissions and cost inventory of power generation technologies in Singapore. *Renew. Sustain. Energy Rev.* **2007**, *11*, 702–715. [CrossRef]

21. Fantin, V.; Giuliano, A.; Manfredi, M.; Ottaviano, G.; Stefanova, M.; Masoni, P. Environmental assessment of electricity generation from an Italian anaerobic digestion plant. *Biomass Bioenergy* **2015**, *83*, 422–435. [CrossRef]

22. Tomasini-Montenegro, C.; Santoyo-Castelazo, E.; Gujba, H.; Romero, R.J.; Santoyo, E. Life cycle assessment of geothermal power generation technologies: An updated review. *Appl. Therm. Eng.* **2017**, *114*, 1119–1136. [CrossRef]

23. Turconi, R.; O'Dwyer, C.; Flynn, D.; Astrup, T. Emissions from cycling of thermal power plants in electricity systems with high penetration of wind power: Life cycle assessment for Ireland. *Appl. Energy* **2014**, *131*, 1–8. [CrossRef]

24. Ehtiwesh, I.A.S.; Coelho, M.C.; Sousa, A.C.M. Exergetic and environmental life cycle assessment analysis of concentrated solar power plants. *Renew. Sustain. Energy Rev.* **2016**, *56*, 145–155. [CrossRef]

25. Peric, M.; Komatina, M.; Bugarski, B.; Antonijevic, D. Best Practices of Biomass Energy Life Cycle Assessment and Possible Applications in Serbia. *Croat. J. For. Eng.* **2016**, *37*, 375–390.

26. Gu, H.M.; Bergman, R. Cradle-to-grave life cycle assessment of syngas electricity from woody biomass residues. *Wood Fiber Sci.* **2017**, *49*, 177–192.

27. Honus, S.; Kumagai, S.; Němček, O.; Yoshioka, T. Replacing conventional fuels in USA, Europe, and UK with plastic pyrolysis gases—Part I: Experiments and graphical interchangeability methods. *Energy Convers. Manag.* **2016**, *126*, 1118–1127. [CrossRef]

28. Honus, S.; Kumagai, S.; Yoshioka, T. Replacing conventional fuels in USA, Europe, and UK with plastic pyrolysis gases—Part II: Multi-index interchangeability methods. *Energy Convers. Manag.* **2016**, *126*, 1128–1145. [CrossRef]

29. Liang, X.; Wang, Z.; Zhou, Z.; Huang, Z.; Zhou, J.; Cen, K. Up-to-date life cycle assessment and comparison study of clean coal power generation technologies in China. *J. Clean. Prod.* **2013**, *39*, 24–31. [CrossRef]

30. Burchart-Korol, D.; Korol, J.; Czaplicka-Kolarz, K. Life cycle assessment of heat production from underground coal gasification. *Int. J. Life Cycle Assess.* **2016**, *21*, 1391–1403. [CrossRef]

31. Hyder, Z.; Ripepi, N.S.; Karmis, M.E. A life cycle comparison of greenhouse emissions for power generation from coal mining and underground coal gasification. *Mitig. Adapt. Strateg. Glob. Chang.* **2016**, *21*, 515–546. [CrossRef]

32. Koci, V.; Trecakova, T. Mixed municipal waste management in the Czech Republic from the point of view of the LCA method. *Int. J. Life Cycle Assess.* **2011**, *16*, 113–124. [CrossRef]

33. Luňáčková, P.; Průša, J.; Janda, K. The merit order effect of Czech photovoltaic plants. *Energy Policy* **2017**, *106*, 138–147. [CrossRef]

34. Europe Commission. *EU Reference Scenario 2016—Energy, Transport and GHG Emissions—Trends to 2050*; The European Commission Report; Europe Commission: Brussels, Belgium, 2016.

35. International Organization for Standardization (ISO). 14044:2006—Environmental Management. Life Cycle Assessment. Requirements and Guidelines. Available online: https://www.saiglobal.com/pdftemp/previews/osh/iso/updates2006/wk26/iso_14044-2006.pdf (accessed on 1 July 2006).

36. Intergovernmental Panel on Climate Change. IPCC Fifth Assessment Report. The Physical Science Basis. 2007. Available online: http://www.ipcc.ch (accessed on 15 November 2017).

37. Energy Policies of IEA Countries—Czech Republic 2016. Review The International Energy Agency. Available online: http://www.iea.org (accessed on 8 November 2017).

38. The Voice of Coal in Europe. Available online: https://euracoal.eu/info/country-profiles/czech-republic/ (accessed on 8 January 2018).

39. State Energy Policy of the Czech Republic. Available online: https://www.mzp.cz/C125750E003B698B/en/climate_energy/$FILE/OEOK-State_Energy_Policy-20160310.pdf (accessed on 24 November 2017).

40. Ministry of Agriculture of the Czech Republic. *Action Plan for Biomass in the Czech Republic for the Period 2012–2020*; Ministry of Agriculture: Prague, Czech Republic, 2016.

41. Czech Republic—Energy System Overview. Available online: http://www.iea.org/media/countries/CzechRepublic.pdf (accessed on 15 November 2017).

environments

MDPI

Article

Environmental Impact of Small Hydro Power Plant—A Case Study

Martina Zeleňáková [1,*], Rastislav Fijko [1], Daniel Constantin Diaconu [2] and Iveta Remeňáková [1]

[1]   Department of Environmental Engineering, Technical University of Košice, 042 00 Košice, Slovakia; rastislav.fijko@tuke.sk (R.F.); iva.remenakova@gmail.com (I.R.)
[2]   Department of Meteorology and Hydrology, University of Bucharest, 010041 Bucharest, Romania; ddcwater@yahoo.com
*   Correspondence: martina.zelenakova@tuke.sk; Tel.: +421-55-602-4270

Received: 30 November 2017; Accepted: 8 January 2018; Published: 10 January 2018

Abstract: Currently an international topic—not only among the members of the European Union—is the use of renewable energy, such as hydro power. The subject of this paper is the environmental impact assessment of the small hydropower (SHP) plant. The paper identifies the environmental impacts of an SHP plant in Spišské Bystré, Slovakia. It also assesses the alternatives to a specific hydraulic structure by quantitative evaluation from the point of view of character of the impacts, their significance, and their duration. The conclusion of the work includes the selection of the optimal alternative of the assessed construction and proposes measurements to reduce the negative impacts. The benefit of this paper is in highlighting the importance of assessing the impact of construction on the environment in the planning phase. Eliminating the negative impacts of the construction on the environment is much more challenging than the implementation of preventive measures, and it is therefore necessary to assess at the planning phase how the construction and operation of the proposed activities impact the environment.

Keywords: environmental impact assessment; hydro power plant; matrix of impacts

1. Introduction

Like the power industry, water management is not a sector per se, but it does secure access to water for all other sectors and for society as a whole according to need. However, unlike energy there are no alternative sources of water, and that is why for several years now water has been considered as a strategic raw material. The assessment of water resources involves establishing the amount, quality, and availability by evaluation of the possibilities of sustaining their development, management, and control. Water is used for cooling, shipping, and washing as a solvent, and also sometimes is found in the ingredients of finished products. A large amount of water is needed for refrigeration equipment. Volumes of industrial water are completely different in individual industrial sectors and also in different types of production, depending on the technology of the production process. Again, this depends on the climatic conditions, because the use of industrial water usually seems to be significantly smaller in northern areas than in southern regions, where the air temperature is higher. Developing the use of industrial water is one of the main reasons for water pollution in the world today. This is explained by the fact that in various countries, industrial growth has greatly increased and exaggerated the proportions of waste being released as waste water into watercourses, predominately untreated or only partially purified. In the battle with such pollution problems, many countries have approved energy measures for reducing the use and release of industrial waters. Since the 1970s and 1980s, a tendency towards stabilization and even a drop in the demand for industrial waters has been seen. It is expected that in many countries in the future, the trend will be a downward one due to a larger use of systems for supplying circulating water, and many industrial branches will aim at dry technologies without water usage.

The issue of greenhouse gas emissions in Europe is becoming important. European Union countries are continuously working to reduce these emissions. They have created a program for the use of renewable energy sources, committing to a 20% reduction of emissions by 2020. Working towards such a goal means that it is necessary to avoid deforestation, use new technologies, and use renewable energy sources—either geothermal, solar, wind, or hydropower [1]. In Slovakia, hydropower is the most common source of renewable energy used to produce electricity. Based on hydropower potential available for electricity generation at 7361 GWh per year, the current use is 57.5%. The share of larger hydropower for electricity produced in 2002 was 92% and the proportion of small hydropower (SHP) only 8%. The utilization of hydro power plants—particularly SHPs—in Slovakia for electricity generation is of prime importance to the economy. Small hydropower plants' utilization of the total available potential is 16% (1220 GWh), while the current total utilization is 24.5% (284.1 GWh) of the total available potential of SHP in Slovakia [2]. Small hydropower (SHP) plants have an installed capacity of 1–10 MW, and their impact on the environment is subject to assessment under Annex 1 of Act no. 24/2006 Coll. the impact assessment on the environment, as amended in the Slovak Republic. Installations for hydroelectric energy production are also subject to an environmental impact assessment under the European Directive 2014/52/EC. The necessity of more comprehensive standards for the impact assessment and the governance of small hydropower projects was proved, for example, by Kibler and Tullos [3]. They investigated the cumulative biophysical effects of small and large hydropower dams in China's Nu River, and they revealed that biophysical impacts of small hydropower may exceed those of large hydropower—particularly with regard to habitat and hydrologic change. Standards for SHP plants' impact assessment are necessary to encourage low-impact energy development. This is also a contribution of the presented paper. The environmental implications of small and large hydropower projects were also studied by Henning et al. [4,5] and Ferreira et al. [6]. Mayor et al. [7] assessed the differential contributions to the regional energy and water security of large- and small-scale hydropower deployment in the Spanish Duero basin. Results of their study showed greater impacts of SHP, mainly as a result of cumulative effects cascading along the rivers system.

Directive 2001/77/EC of the European Parliament and of the Council on the promotion of electricity produced from renewable energy sources in the internal electricity market and Directive 2003/54/EC of the European Parliament and of the Council have stability rules commune in terms of obtaining and distributing electricity. The Commission Communication of 10 January 2007 entitled *"Roadmap for renewable energy—Renewable energies in the 21st century: building a more sustainable future"* has demonstrated that a 20% target for the global share of renewable energy would be an achievable target and that a framework that includes binding targets should provide the business community with the long-term stability needed to make rational and sustainable renewable energy investments to reduce dependence on imported fossil fuels and increase the use of new energy technologies. The Framework Directive 2009/28/EC on the promotion of the use of energy from renewable sources aims at developing the local and regional electricity market in order to reduce greenhouse gas emissions. The Renewable Energy Directive establishes an overall policy for the production and promotion of energy from renewable sources in the EU Member States. It requires the EU to fulfil at least 20% of its total energy needs with renewables by 2020—to be achieved through the attainment of individual national targets. On 30 November 2016, the Commission published a proposal for a revised Renewable Energy Directive to ensure that the target of at least 27% renewables in the final energy consumption in the EU by 2030 is met. The Directive specifies national renewable energy targets for each country (from a low of 10% in Malta to a high of 49% in Sweden), taking into account its starting point and overall potential for renewables.

However, all of these initiatives are in conflict with the provisions of the Framework Directive 2000/60/EC, according to which strict rules are imposed to reduce the hydromorphological alteration of watercourses, but also the evaluation of eco-systems. In Romania, the development of investments in micro-hydro power plants was supported, but in contradiction the environmental provisions were enforced, imposing major restrictions. So, in 2015, the European Commission launched the infringement procedure against Romania due to micro-hydropower projects in the Fagaras Mountains. A similar situation has occurred in Slovakia.

Assessment of the impact of the project on the environment is considered as a tool that minimizes the implementation of activities which could in any way negatively interfere with the environment and at the same time allows choosing the optimal solution from the proposed alternatives of the project implementation—the alternative with the smallest negative impact of a proposed activity on the environment. At present, several authors are devoted to the issue of environmental impact assessment in Slovakia [8–11]; in the Czech Republic [12]; in Poland [13]; and in Romania [14,15].

Environmental impact assessment (EIA) procedures for public and private projects that are likely to have significant effects on the environment in Slovakia have been in place since the adoption of the EIA Act in 1994. In 2006, a new EIA Act was approved, and EIA procedures began to be applied to buildings under the 2006 Planning Act. Law no. 24/2006 Coll. on the assessment of impacts on the environment and on amendments to certain laws, which entered into force on 1 February 2006 to regulate all EIA process in the Slovak Republic. It implements Directive 2014/52/EC of the European Parliament and the Council amending the previous Directive 2011/92/EC on the assessment of the effects of certain public and private projects on the environment. EIA is a tool for decision making, with the final aim of the sustainable development of society.

According to Law no. 24/2006 Coll. the assessment of impacts on the environment and on amendments in the Slovak Republic, the "Industrial installations for the production of electricity from water power" hydropower plants from 5 MW to 50 MW are under a screening procedure, and hydropower plants producing more than 50 MW are under compulsory assessment.

This paper also briefly presents a case study: technical and technological solutions of three variants of the selected engineering construction of a small hydroelectric power plant situated in Spišské Bystré, in Slovakia. The assessment was devoted to direct impacts of the proposed activity, characteristic of the current state of the environment in the affected area, the assessment of the expected impacts of the proposed activity on the environment, and estimation of their importance using the point method. For comparison of the variants of small hydroelectric power plants, the matrix method is applied. The purpose of this comparison is to select the optimal variant of the proposed action, and proposes measures to prevent, eliminate, minimize, and offset the impacts of the proposed activity on the environment.

2. Materials and Methods

Many methods have been introduced over the last 50 years to meet the different requirements of environmental impact assessment studies. Mentioned methods are explained, for example, in [16,17]. There is a need for a general and thorough approach to justifying, explaining, demonstrating, implementing, sampling, using, and creating real skills in analysis in any area of human society [18,19]. Most management decisions are concerned with the future; however, the future is usually uncertain [20–25]. The uses of risk identification, analysis, and assessment in relation to the environment have broadened considerably in recent years.

Guidelines of the European Commission [22] provide information on approaches that were selected from case studies and literature survey. These include scoping and screening techniques which predict the magnitude and significance of impacts and attempt to quantify them based on their intensity, frequency, duration, and character. Scoping and impact identification methods include:

- Network and analysis
- Consultation and questionnaires
- Checklists

Evaluation and screening methods include:

- Modeling
- Comparative methods

Techniques include:

- Matrices
- Expert opinion.

The EIA process involves a combination of approaches [12]:

- Identification and definition of the impact;
- Analysis of the impact associated;
- Determining the significance of the impact.

It is expected that EIA will continue to act as an effective tool to prevent the application of investments not only in Slovakia which by their degree of environmental damage vastly outweigh their benefits [23–25].

In this paper, the impact matrix has been used for the environmental impact assessment of small hydropower plants, and presents an overview, distribution, and classification of the impact of the projects on the environment by different criteria for the purposes of the evaluation. In addition, it also highlights the identification and assessment of the expected impacts of the construction on the environment. The presented impact matrix combines qualitative and quantitative methods: verbal statements which were transformed into the numerical values presented in Table 1. This assessment requires special attention and sensitive work with verbal and numerical scales. It used indicator values. This method consists of only a very approximate method, where by its value an indicator may represent a description of the analyzed problem.

The assessment was done by seven experts—the authors and three more people—the experts working in the field of SHP plant design and/or the assessment of environmental impacts of SHP plants, and one of them is working in the landscape ecology—the nature protection. They used the brainstorming method. They consulted the selection of the criteria and their impacts at the personal meetings as well as by e-mail communication.

The proposed EIA methodology involves a combination of approaches:

- Establishing the context

  ○ Characteristics of the current state of the environment in the affected area
  ○ Brief description of alternatives of the proposed activity (*A*0, *A*1, *A*2)

- Evaluation of impacts

  ○ The character of the impacts
  ○ The significance of the impacts
  ○ The duration of the impacts

- Quantification of impacts and
- Comparison of alternatives.

The use of proper EIA methodologies and procedures can help the decision-makers to manage proper activities based on qualified decisions [26].

In the following, the case study is presented.

Table 1. Quantification of impacts and comparison of alternatives.

Impact on	Alternative 0				Alternative 1				Alternative 2			
	CH0	S0	D0	CH0 × S0 × D0	CH1	S1	D1	CH1 × S1 × D1	CH2	S2	D2	CH2 × S2 × D2
population:												
noise	0			0	−1	2	1	−2	−1	2	1	−2
vibrations	0			0	−1	2	0.5	−1	−1	2	0.5	−1
dust	0			0	−1	3	0.5	−1.5	−1	3	0.5	−1.5
quality of life	0			0	−1	2	0.5	−1	−1	2	0.5	−1
economy	−1	2	1	−2	1	3	1	3	1	2	1	2
tourism, recreation	0			0	1	2	1	2	1	2	1	2
sport activities	0			0	1	2	0.5	1	1	2	0.5	1
water conditions:												
surface water flowing	−1	2	1	−2	1	2	1	2	1	2	1	2
surface water standing	−1	2	1	−2	1	3	1	3	1	3	1	3
ground water in inundation	−1	2	1	−2	1	3	1	3	1	3	1	3
ground water in protected area	−1	2	1	−2	1	2	1	2	1	2	1	2
soil:												
land occupation	0			0	−1	1	1	−1	−1	1	1	−1
water regime of soil	−1	2	1	−2	1	2	1	2	1	2	1	2
soil erosion	−1	2	1	−2	1	3	1	3	1	3	1	3
fauna and flora and their biotopes:												
fauna-mammals	1	2	1	2	−1	2	1	−2	−1	2	1	−2
fauna-birds	1	1	1	1	−1	1	1	−1	−1	1	1	−1
fauna-ichthyofauna	1	2	1	2	−1	3	1	−3	−1	1	1	−1
fauna-amphibians	1	2	1	2	−1	3	1	−3	−1	1	1	−1
flora-at construction site	0			0	−1	3	0.5	−1.5	−1	2	0.5	−1
flora-at backwater	0			0	−1	2	1	−2	−1	1	1	−1
landscape:												
structure	0			0	−1	1	1	−1	−1	1	1	−1
using	0			0	1	2	1	2	1	2	1	2
scenery	0			0	−1	1	1	−1	−1	1	1	−1
the protected areas and their protective zones:												
protected areas	0			0	−1	1	1	−1	−1	1	1	−1
water protected areas	0			0	−1	1	1	−1	−1	1	1	−1
the territorial system of ecological stability	0			0	−1	3	1	−3	−1	2	1	−2
urban areas, land use:												
urban areas	0			0	1	3	1	3	1	3	1	3
land use	0			0	1	3	1	3	1	3	1	3
air:												
air quality	0			0	−1	1	1					
concentrations of emissions	0			0	−1	1	1					
SUM				−7				1				7.5

Study Area

The selected site for the proposed activity—construction of small hydropower plant—is located in the village Spišské Bystré, district of Poprad, Eastern Slovakia (see Figure 1). The study area is close to the High Tatras Mountains, where the High Tatras National Park is located (20 km) and near the Slovak Paradise protection area (10 km), but the selected site is actually out of any protected area. The municipality is located at an altitude of 674 m, has a population of 2394 and an area of 3787 ha. Bystrá creek is a right-bank tributary of the Hornád, which belongs to Danube River Basin and has a length of 17 km. The proposed SHP plant is designed as a run-of-river plant in river kilometer 4.0. The annual discharge of Bystrá creek is 0.42 m^3/s. The discharge of 100-years return period in the site is 60 m^3/s. Currently, the creek in that area has the character of unregulated water flow with an irregular trapezoidal profile width from 2.0 m to 6.0 m in the bottom and from 5.0 m to 10.0 m in water level. The proposal of a small hydropower plant includes a regulation of the river bed, which consists mainly of fortifications of the channel cross-section.

Three alternatives of the proposed activity were assessed (Figure 2):

- Alternative 0—the present state of the environment, no SHP plant will be constructed;
- Alternative 1—construction of SHP plant;
- Alternative 2—construction of SHP plant with bypass fish pass.

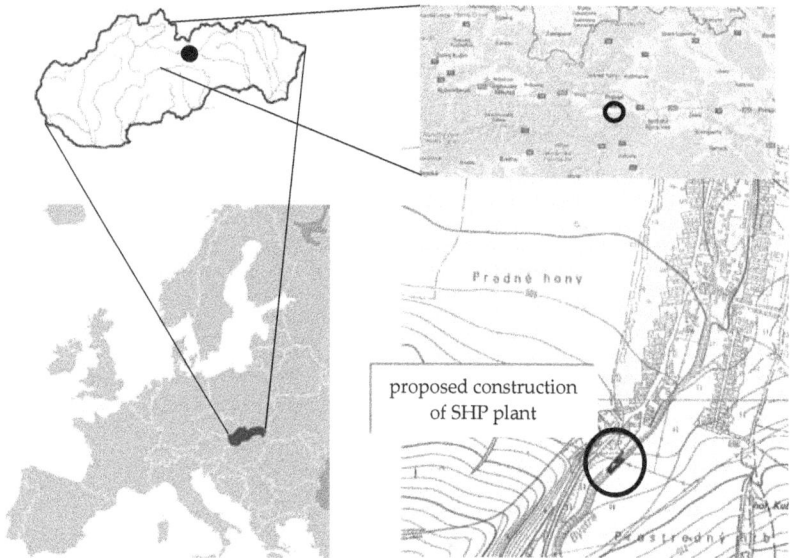

Figure 1. Location of the study area. SHP: small hydropower.

Figure 2. *Cont.*

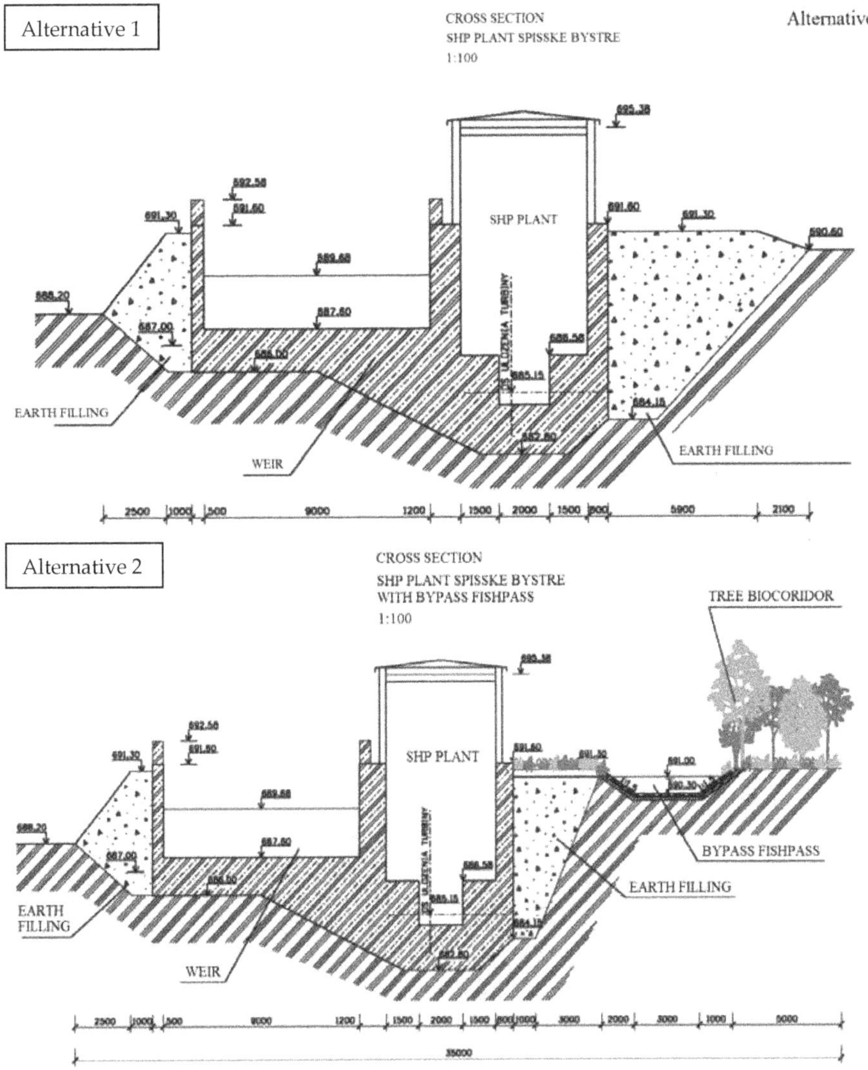

Figure 2. Alternatives of the proposed activity.

Impacts on the environment are reflected by the effects of the water project, which in this case is the small hydro power plant. From the identification of impacts and their influence on individual components of the environment of the study area by the detection matrix, it is obvious that the most significant negative impacts during the construction activity will be from construction machinery, accompanied by noise, emissions, and dust. These stressors negatively affect habitats, climate, population and other components of the environment. The construction will also affect the soil layers as well as the quality of surface water. Direct negative impact on the environment during the operation of the SHP plant is not expected. For a comprehensive assessment of the expected impact in terms of its significance, nature, and duration, it is recommended to use the quantitative evaluation method. The assessment of the expected impacts of the proposed activity on the environment is presented as follows.

3. Results

This chapter includes the comparison of alternatives of the proposed activity (SHP plant in Spišské Bystré) and the selection of the optimal alternative, including a comparison with zero alternative (Alternative 0, Alternative 1, Alternative 2). The first step of the impact assessment of the proposed activity on the environment is the identification of the impacts on the partial components of the environment. When developing criteria and determining their importance, we placed emphasis on the nature, extent, and duration of the effects. We have assigned values to individual consequences according to the proposed scale:

- The character of the impacts (CH):

  - − negative,
  - 0 no impact,
  - + positive,

- The significance of the impacts (S):

  - 1 insignificant,
  - 2 significant,
  - 3 very significant,

- The duration of the impacts (D):

  - 0.5 short-term,
  - 1 long-term.

Table 1 identifies the impacts on individual components of the environment.

The values of the nature of the impacts are added according to the above-proposed scale. We have assessed only the impacts envisaged occurring during the construction and operation of the SHP plant.

The next step in the EIA process is the selection of the optimal alternative by assessing the character, significance, and duration of the impacts undertaken by quantitative method (Table 1). Impacts' nature have been counted separately for each alternative of the proposed activity as a sum of points that are product of multiplying the character, the significance, and the duration of the impacts. The alternative that reached the highest positive value can be considered as optimal.

The highest value of 7.5 is according to Alternative 2: SHP plant Spišské Bystré with a bypass fishpass; therefore, it can be considered from a comprehensive assessment of the environmental impact as an optimal variant, with the least negative impacts to the environment (although this alternative is the costliest). During construction, it is necessary to pay attention to the measures that reduce and respectively mitigate the adverse impact on the environment, including the health of the population. Measures need to be designed to prevent, eliminate, minimize, and compensate the negative impacts.

4. Discussion

The objective of this paper is the analysis and evaluation of the environmental impacts of small hydro power plants by use of the matrix of impacts. The identification and evaluation of the environmental impacts include the health and social impacts to the population in the study area. The aim is the selection of the optimal variant of the proposed activity, using quantitative evaluation. In practice, there are many procedures and methods which can be used to identify and evaluate these impacts. In order to identify environmental impacts of the water structure, in accordance with valid legislation, we used a survey matrix method by which we specified the impacts of the activity on the components of the environment. This method increases the transparency and precision of the evaluation process, and also satisfies the requirements of the environmental impact assessment procedure. This approach can be applied to other infrastructure projects. Comparison of alternatives and selection of the optimal variant are implemented based on selected criteria and foreseen impacts of the activity on the environment. This method is adequate for the assessment of the proposed environmental impacts of SHP plants, although to confirm the results, the other (the alternative) method could be also used. Comparison of the variants revealed that Alternative 2—construction of SHP plant with bypass fishpass can be considered as an optimal variant because it reached the highest positive values. Not only from an economic point of view (which is the use of renewable energy sources), but also from an ecological point of view that involves the reduction of greenhouse gas emissions, the proposal of the optimal variant is beneficial to the concerned area. The benefit of the study is in pointing out the importance of environmental impact assessment of the construction before requesting a permit for the construction. Eliminating the negative effects of construction on the environment is far more demanding than the implementation of preventive measures, and it is therefore necessary to assess how the construction will affect operation of the proposed activity in the area. The use of the EIA process can help the decision-makers to select proper mitigation measures based on qualified decisions. The proposed activity in Spišské Bystré, Slovakia is used as a case study to exemplify the methodology.

5. Conclusions

Environmental impact assessment (EIA) is an important process prior to approval of the proposed activity. It can provide essential information about the foreseeable impacts of the investment plan on the environment. The assessment of the potential impacts on the environment is the most important stage in the EIA process. Environmental assessment is based on the technical description of the project, as well as prediction and evaluation of the impacts on the environmental components. From the point of view of environmental requirements for construction, the negative impact on the environment is minimized in the preparatory phase of the project by analyzing and assessing the impact of the construction on the environment, thus avoiding an increase in costs due to unforeseen impacts during the construction phase.

The knowledge of tools to assess interaction between humans, natural resources, and water projects is developed, distributed, and used with the aim of mitigating adverse impacts and remediating the environment. The approach has an original solution concept.

Acknowledgments: This work was supported by VEGA project 1/0609/14.

Author Contributions: Martina Zeleňáková prepared and edited the manuscript; Rastislav Fijko worked on the materials and methods section; Daniel Constantin Diaconu prepared the introduction section; Iveta Remeňáková worked on the drawings. All authors did the evaluation of the environmental impacts and approved the final manuscript.

Conflicts of Interest: The authors declare no conflicts of interest.

References

1. European Commission—Renewable Resources of Energy. Available online: http://www.ec.europa.eu/news/energy/120608_sk.htm (accessed on 2 March 2015).
2. Hydropower Energy. Available online: http://www.oze.stuba.sk/oze/vodna-energia/ (accessed on 2 March 2015).
3. Kibler, K.K.; Tullos, D.D. Cumulative biophysical impact of small and large hydropower development in Nu River, China. *Water Resour. Res.* **2013**, *49*, 3104–3118. [CrossRef]
4. Mayor, B.; Rodríguez-Muñoz, I.; Villarroya, F.; Montero, E.; López-Gunn, E. The role of large and small scale hydropower for energy and water security in the Spanish Duero Basin. *Sustainability* **2017**, *9*, 1807. [CrossRef]

5. Hennig, T.; Wang, W.; Feng, Y.; Ou, X.; He, D. Review of Yunnan's hydropower development. Comparing small and large hydropower projects regarding their environmental implications and socio-economic consequences. *Renew. Sustain. Energy Rev.* **2013**, 27, 585–595. [CrossRef]
6. Hennig, T.; Wang, W.; Magee, D.; He, D. Yunnan's fast-paced large hydropower development: A powershed-based approach to critically assessing generation and consumption paradigms. *Water* **2016**, 8, 476. [CrossRef]
7. Ferreira, J.H.I.; Camacho, J.R.; Malagoli, J.A.; Camargo Guimarães, S., Jr. Assessment of the potential of small hydropower development in Brazil. *Renew. Sustain. Energy Rev.* **2016**, 56, 380–387. [CrossRef]
8. Zeleňáková, M.; Zvijáková, L. *Using Risk Analysis for Flood Protection Assessment*; Springer: Cham, Switzerland, 2017; p. 128. ISBN 978-3-319-52150-3.
9. Pavlíčková, K.; Kozová, M.; Miklošovičová, A.; Zarnovičan, H.; Barancok, P.; Luciak, M. Environmental impact assessment. In *Textbook for Students of Master's Studies*, 1st ed.; Comenius University in Bratislava: Bratislava, Slovakia, 2009.
10. Kočická, E. Environmental Impact Assessment (in Term of Theory and Praxis). Available online: https://stary.tuzvo.sk/files/FEE/dekanat_fee/11_Kocicka_AFE.pdf (accessed on 30 November 2017).
11. Majerník, M.; Bosák, M. *Environmental Impact Assessment*; Technical University: Košice, Slovakia, 2003.
12. Říha, J. *Environmental Impact Assessment of Investments. Multicriteria Analysis and EIA*; Academia: Prague, Czech Republic, 2001.
13. Gałaś, S.; Gałaś, A.; Zeleňáková, M.; Zvijáková, l.; Fialová, J.; Kubíčková, H.; Házi, J. *Comparing the Phase of Screening in the Fields of Tourism and Recreation Water Management and Mining in the V4 Countries*; AGH University of Science and Technology Press: Krakow, Poland, 2014.
14. Rojanschi, V.; Bran, F.; Diaconu, G. *Protection and Environmental Engineering*; Editura Economică: Bucuresti, Romania, 1997; p. 368.
15. Iojă, I.C. *Analysis and Evaluation of the Environmental Situation*; Editura Economică: Bucuresti, Romania, 2013; p. 183.
16. Canter, L.W. Methods for effective environmental impact assessment. Environmental methods review. In *Retooling Impact Assessment for the New Century*; The Press Club: Fagro, ND, USA, 1998; pp. 58–68.
17. Canter, L.W. Environmental Impact Assessment. In *Environmental Engineers' Handbook*; CRC Press: Boca Raton, FL, USA, 1999.
18. Bujoreanu, I.N. Risk analysis series part one—Why risk analysis? *J. Def. Resour. Manag.* **2012**, 3, 139–144.
19. Romanescu, G.; Miftode, D.; Mihu-Pintilie, A.; Stoleriu, C.C.; Sandu, I. Water quality analysis in mountain freshwater: Poiana Uzului reservoir in the eastern carpathians. *Rev. Chim.* **2016**, 67, 2318–2326.
20. Costanza, R.; Daly, H.E. Natural capital and sustainable development. *Conserv. Biol.* **1992**, 6, 37–46. [CrossRef]
21. Welsh, T. Full Monte—The Better Approach to Schedule Risk Analysis User Guide. Available online: http://www.barbecana.com/_downloads/Full%20Monte%202017%20User%20Guide.pdf (accessed on 2 December 2016).
22. Walker, L.J.; Johnston, J. Guidelines for the assessment of indirect and cumulative impacts as well as impact interactions. In *Environment, Nuclear Safety & Civil Protection*; The Publications Office of the European Union: Luxembourg, 1999; p. 170.
23. Zvijáková, L.; Zeleňáková, M.; Purcz, P. Evaluation of environmental impact assessment effectiveness in Slovakia. *Impact Assess. Proj. Apprais.* **2014**, 32, 150–161. [CrossRef]
24. Zvijáková, L.; Zeleňáková, M. *Risk Analysis in the Process of Environmental Impact Assessment of Flood Protection Objects*; Leges: Prague, Czech Republic, 2015; p. 255.
25. Tkáč, Š.; Vranayová, Z. The use of the water element in the energetics of micro-urban development in Slovak Republic and Taiwan R.O.C. *Pollack Periodica* **2014**, 9, 101–112. [CrossRef]
26. Shah, A.; Salimullah, K.; Sha, M.H.; Razaulkah, K.; Jan, I.F. Environmental impact assessment (EIA) of infrastructure development projects in developing countries. *Int. J. Sustain. Dev.* **2010**, 1, 47–54.

environments

MDPI

Article

Selection of the Best Alternative of Heating System by Environmental Impact Assessment—Case Study

Vlasta Ondrejka Harbulakova [1,*], Martina Zelenakova [1], Pavol Purcz [2] and Adrian Olejnik [1]

[1] Department of Environmental Engineering, Faculty of Civil Engineering, Technical University of Kosice, Vysokoskolska 4, 042 00 Kosice, Slovakia; martina.zelenakova@tuke.sk (M.Z.); adrianolejnik91@gmail.com (A.O.)

[2] Department of Applied Mathematics, Faculty of Civil Engineering, Technical University of Kosice, Vysokoskolska 4, 042 00 Kosice, Slovakia; pavol.purcz@tuke.sk

* Correspondence: vlasta.harbulakova@tuke.sk; Tel.: +421-55-602-4269

Received: 9 December 2017; Accepted: 24 January 2018; Published: 26 January 2018

Abstract: The Framework Directive 2009/28/EC on the promotion of the use of energy from renewable sources aims at developing the local and regional electricity market in order to reduce greenhouse gas emissions. A comparison study of the proposed activities of construction of a new biomass-fired power plant or reconstruction of an old one-gas power plant in town located in eastern Slovakia is presented in this paper. The method of the index coefficient was used for choosing the best alternatives. Multicriteria analysis proved that the construction of biomass-fired power plant is the most suitable solution chosen from three assessed variants (no activity is implemented, biomass power plant and modernized gas boiler).

Keywords: biomass-fired power plant; ecological criteria; economic criteria; multicriteria analysis; technical criteria and technological criteria

1. Introduction

The Cardiff summit in 1989 created the platform of coordinated action aimed at protecting the environment. The European Commission has progressively focused its attention on the development and integration of environmental aspects into the sectoral policies of transport, energy, industry, agriculture, industry, internal economic policy and fisheries. The first step was decision-making of the first integration strategy in the energy sector adoption in 1999, which was modified in 2001 and presented in Gothenburg, Sweden, before the European Council [1,2].

Directive 2001/77/EC [3] of the European Parliament and of the Council of 27 September 2001 on the promotion of electricity produced from renewable energy sources in the internal electricity market and Directive 2003/54/EC [4] of the European Parliament and of the Council of 26 June 2003 have stability rules commune in terms of obtaining and distributing electricity. Another important document presented by the European Commission was "Green Paper on a secure, competitive and sustainable energy for Europe" [5], which was released in 2006. The aim of the EC was to create an integrated energy policy in Europe. The Commission Communication [6] of 10 January 2007 entitled "Roadmap for renewable energy—Renewable energies in the 21st century: building a more sustainable future" has demonstrated that a 20% target for the global share of renewable energy would be achievable targets and that a framework that includes binding targets should provide the business community with the long-term stability needed to make rational, sustainable renewable energy investments to reduce dependence on imported fossil fuels and increase the use of new energy technologies.

In December 2008, a wide range of measures in the European Union (EU) was adopted, which were aimed at reducing the impact of the EU states' activities on global warming and also reducing negative effects on the global climate, while ensuring adequate and reliable energy supply. The Framework

Directive 2009/28/EC [7] on the promotion of the use of energy from renewable sources aims at developing the local and regional electricity market in order to reduce greenhouse gas emissions.

Energy is one of the sectors that pollute and harm the environment the most. The alignment of energy and environmental conditions is at the moment one of the most serious strategic challenges to addressing global environmental problems. Therefore, the development of energy must strictly respect the principles of sustainable development. With a view to the future, the reduction of the negative effects of energy on the environment in Slovakia can be achieved by promoting the use of renewable energy sources as well as by promoting energy-saving measures [3].

From the point of view of sustainable development, the transition from the use of non-renewable energy sources to the use of renewable energy sources is necessary. To achieve this goal, it is necessary to change habits, practices and technologies not only in production but also in consumption. Total energy consumption is one of the main determinants of the impact of energy on the environment. Therefore, it is necessary to harmonize the relationship between energy and the environment through the introduction of technologies using renewable energy sources. All energy sources must be used not only with respect to the environment but also with respect to human health [8]. There are many Environmental Impact Assessments (EIA) of the thermal power plant worldwide. The energy strategy of the EU has become one of the most important factors influencing the development not only in the polity of the member states. Different mitigation measures for the control of pollution caused by thermal power as well as some new technologies are described in the paper by authors from India [9]. The Environmental Impact Assessment Report for the Tanda thermal power plant (in New Delhi) presents that the adverse environmental impacts due to construction and operation of thermal power can be mitigated to an acceptable level by implementation of various measures for mitigation [10]. In a paper by Pokale [11], the EIA of the thermal power plant and its effect, as well as a cost-benefit analysis, are presented and discussed. Studies not only from Italy [12], Bangladesh [13], China [14,15], Spain [16] or Turkey [17] confirm that the countries try to make every effort to respect the principles of sustainable development. Moridi et al. (2015) [18] presented a Multi-Criteria Decision-Making analysis method (MCDM) with the aim to select and prioritize appropriate technologies based on key criteria including cost, design, maintainability, and size and filtration efficiency in their paper. They were searched for optimal technologies for the treatment of air pollutants emitting from industrial complexes, especially petrochemical industries. EIA is a necessary step also during the early planning stages of environmental structures in order to gain clear insights into the structures' probable impacts with respect to the different components of the total environment. Likewise, the use of appropriate EIA techniques can aid decision-makers in formulating appropriate actions based on informed decisions in light of project urgency and limited resources, which are common constraints in developing countries [19].

This paper is based on legislative and methodological documents relating to the assessment process of environmental impact—Directive 2014/52/EU of the European Parliament and of the Council of 16 April 2014 [20] on the assessment of the effects of certain public and private projects on the environment (only Act in the following) (the Environmental Impact Assessment, or EIA Directive) [20] and exactly Act No. 24/2006 Coll. on Environmental Impact Assessment as amended [21]—national Law in Slovakia (only Act in the following). The purpose of this Act is to ensure the procedure for the overall expert and public assessment of construction, and other activities determined under the Act (see Act Annex) prior to the decisions on the permission thereof under special provisions, and also for the assessment of proposals for certain development policies and generally binding legal directives from the point of view of their presumed effect on the environment.

The main steps of the EIA process in Slovakia according to National Council of the Slovak Republic 2005 are presented in Figure 1 [22].

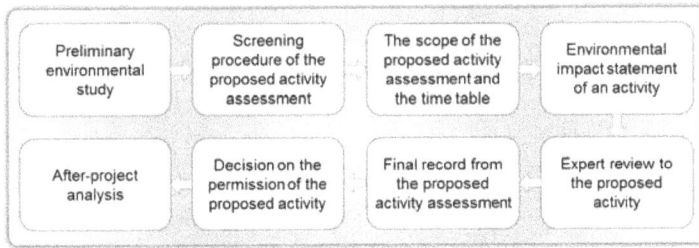

Figure 1. Main steps of the Environmental Impact Assessments (EIA) process in Slovakia (National Council of the Slovak Republic 2005) [22].

The 2006 EIA Act introduced no major changes in EIA procedures; it tightened certain procedural time limits and better delineated EIA responsibilities between the Ministry of the Environment and the regional and district environment offices. The Act was amended many times to respect all requirements of European Union; the last amendment was in June 2017, which was devoted to public participation in the EIA process [22].

There is a direct proportion of energy production and impact of the assessed activity on the environment. Effective and rational use of energy seems to be the best solution how to decrease the negative impact [8].

In the Act, the term "activity" is defined as an operation (structure, facility or others) that by its properties, localization or cooperation with other factors can affect the environment and cultural heritage. To implement such an activity, the permission (approval) or other decision according to specific regulations is necessary.

The Act consists of six parts and sixteen annexes. Annex 8 presents a list of activities subject to environmental impact assessment. It is divided into part A, presenting activities subject to obligatory evaluation, and part B, presenting activities subject to screening.

Variant solutions which are subject to assessment are mostly different in terms of capacity, technology, time of realization, construction costs and maintenance costs. When comparing variants, the result should be in the order of suitability for implementation with regard to human health and the environment. The economic and technical aspects of implementation are the most distinct. The economic aspect is not required by Act 24/2006 [21] to be considered, therefore, in many intentions it is only advisory. Comparing of variations mostly only provides information on the impact of individual variants on the environment and human health [6].

The proposed activity is the construction of a new biomass-fired power plant or reconstruction of an old one-gas power plant in the Trebišov town district in Slovakia. The result is a comparison of the proposed activities with the current state of the area using index coefficient method and finally the selection of the best alternatives of the heating system of Trebišov city.

2. Materials and Methods

2.1. Description of the Study Area

Impact assessment on the environmental compounds of the proposed power plant for heating in Trebišov city in Slovakia (Figure 2) was done according to Act 24/2006 Coll. as amended [21]. Trebišov is situated in the southeast part of Slovakia. It is surrounded by the Slanské Mountains that affect the air circulation. Southwest to northwest winds prevail in this territory. Average annual wind speed in the lowlands of Slovakia is relatively homogeneous. Minimum wind speed is usually connected with situations of temperature inversion, especially in the winter season.

Trebišov district is separated into three climatic areas. Part of the district lies in warm and moderately warm climatic areas and one part lies in cold climatic areas (higher than 800 m a.s.l.) [23].

The study area according to temperature is characterized by two areas. Specifically, part of the district is situated in a warm and dry area with cold winter (January ≤ -3 °C) and part belongs to warm and mildly dry area with cold winter (January ≤ -3 °C).

Long-time average air temperature measured at the Milhostov weather station is 9 °C, and during the vegetative season, the temperature reaches 16.5 °C. The duration of the vegetation period is 200–220 days. The average number of summer days is 67 [23,24].

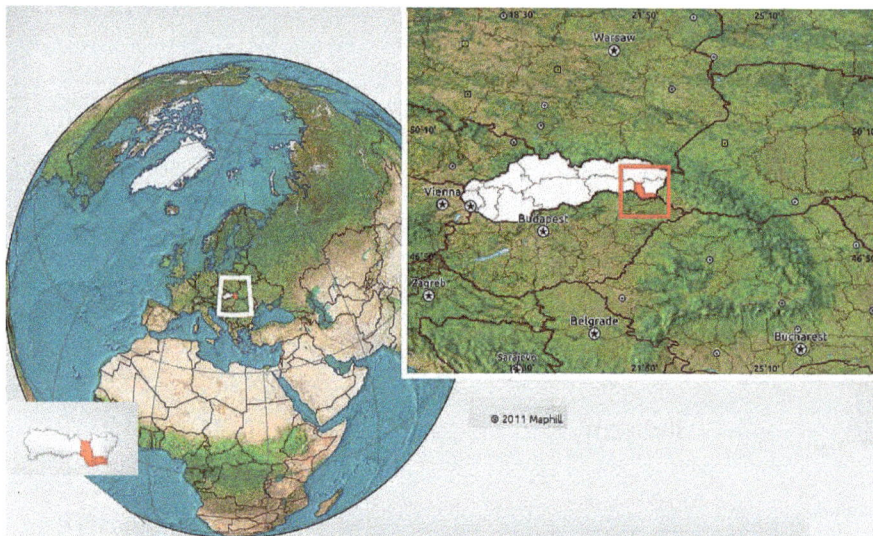

Figure 2. Location of Slovakia and Trebišov district within the European Union (adapted from maphill.com) [25].

The rivers in the Trebišov district are the Ondava River and Latorica River. There are no areas of water bodies which are under the protection. Very important water sources are the Ondava River and Trnávka River, as well as Ruskovský Pond situated in this study area [26]. From the geomorphological division, the study area is located in the Matransko-Slanská area (The Western Carpathians) and in the Eastern Slovakian Lowlands. There are no more significant deposits of ore minerals in the vicinity of Trebišov. There are mainly non-toxic raw materials here as andesite (mainly), which is used primarily as a building material, gravel or paving stone. In particular, the Slanské Hills are covered with andesite volcanoes and stratovolcano.

In the cadastral area of the Trebišov town, the following soils are found: black, brownish and bottomland occurring as subtypes of the black soils and brownish soils [27]. From the agricultural point of view, the most valued are brownish soils. They are more harvested and used for cereal growing. The acidity of the soil is weakly alkaline and has a pH value of 7.3 [27].

2.2. Description of the Project

A brief description of the technical and technological solution is as follows. The central energy source of heating for Trebišov town has two possibilities:

- biomass-fired power plant (alternative 1; wood chips and straw);
- modernization of existing natural gas boiler (alternative 2).

A technological solution of biomass-fired power plant consists of the boiler house (SO 01), a handy storage of straw (SO 02) and technical annex (SO 03). Individual objects are connected structurally and technologically. Objects of power supply are located in the northern part of the plot.

The boiler construction consists of a steel hall with three boilers, one for combustion of wood and two for burning straw. The building object SO 02—a straw storeroom—consists of a single store. There is a firewall designed between the boiler room and the straw storeroom. The building object SO 03—the technical annex—consists of two above-ground floors. Construction of the technical pavilion is built of Porotherm walls. In the technical pavilion, there is the operator, the equipment and the administration of the boiler room [23,24].

2.2.1. Biomass-Fired Power Plant

Wood Chips Combustion

Uncontaminated wood chips will be used as fuel. The preparation of fuel is ensured by cleaving of waste wood (wood-based residue from wood extraction) [23,24]. The fuel will be transported to the boiler room by means of a lorry with a trailer on which two containers are attached with volumes of 2×35 m^3. The cars enter via gate No. 1 to the building and the fuel is emptied to a biomass-wood storage site (building SO 04). It is an uncovered, three-sided (height of 2.5 m) concrete walls bounded space. The entrance of fuel into the landfill, as well as its storage, is illustrated by the red arrows, and the transfer of fuel from the biomass landfill to the charging press is illustrated by the blue arrows in Figure 3. Ash removal from the burning process is illustrated by the green arrows. The task of the fuel store exchanger is to gradually add the accumulated quantity of feedstock (wood chips) to the feed press and through it to the inlet part of the boiler in order to ensure the continuous supply of the boiler to the fuel.

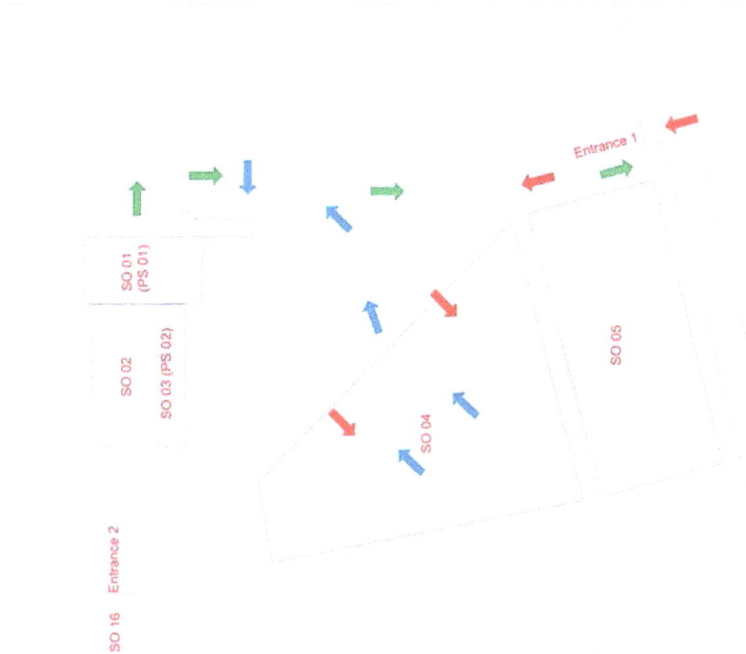

Figure 3. The direction of the fuel handling (source: by authors).

Straw Combustion

As a fuel, the straw used in agriculture as cereal crops will be used. Fuel is provided by packing into large straw bales in the fields.

The fuel will be transported to the boiler room by means of a lorry on which straw bales are stored. The straw is stored on the car; in one layer there are 10 parcels and, in the case of an assumed package with a width of 1200 mm, height 900 mm, length 2200 mm, weight 261 kg, can be placed in three layers. One trailer car carries 60 straw bales for a total weight of 15.66 t [23]. The cars enter via gate No. 1 and the fuel is transported either directly to the straw storeroom (SO 02) or to the straw landfill (SO 05). The placement decision will be based on the fulfilment of both stocks. The unloading is carried out using a forklift truck. The transport of fuel to the boiler house and to the warehouses is illustrated by the red arrows in Figure 4.

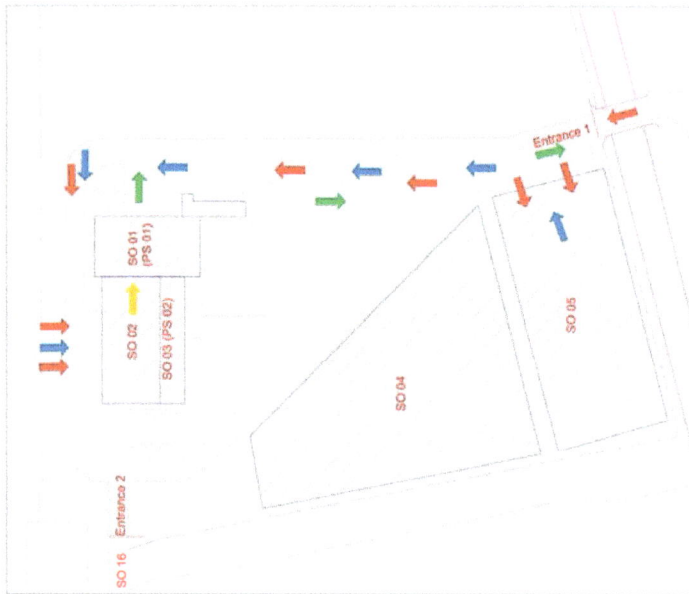

Figure 4. The direction of the straw handling (source: by authors).

The covered straw landfill (SO 05) only serves as a temporary storage to overlap a failure in the fuel supply logistics to the boiler house. In the case of emptying the straw storeroom (object SO 02), straw bales will be taken from the covered straw landfill (object SO 05) using a forklift truck. The transport of fuel from the covered straw landfill to the straw storeroom is illustrated by the blue arrows in Figure 4.

In the case of direct straw removal in the straw storeroom (object SO 02), the packages are stacked on top of a precisely drawn place on the store floor. Storing exactly in place is important for handling straw bundles using an automatic crane. The bridge crane located on the crane track, which will be located on the building of the hall, ensures the transport of the straw package as required by the boiler automation and its placement on the conveyor, which ensures delivery of straw to the boiler. The fuel transportation within the interior spaces is represented by a yellow arrow and ash removal from the burning process is illustrated by the green arrows (Figure 4). Just before entering the straw bales into the boiler, the conveyor ensures that it is rotated to the vertical position. In this position, the straw bundle is cut into three parts by means of the cutting device and transported to the boiler itself, where it burns [23].

2.2.2. Modernization of Existing Natural Gas Boiler

Atmospheric natural gas-fired water tube boilers are used in old central boiler houses. These boilers using the combustion gas transfer the heat to the primary heat exchanger, then it is distributed to the secondary circuit and through pipes to the heated objects. At the secondary circuit is installed a heat meter which measures the heat consumption [23,24,28]. They produce dry combustion products which reach a temperature of 120 °C to 180 °C. Hot combustion products are discharged to the chimney, thereby loss of heat appears. Waste gas contains latent heat which is bound to water vapor resulting from the combustion of natural gas. From the total heat energy gained from natural gas combustion, only 80% is used for water heating. This combustion of natural gas forms a large amount of exhaust gases emitted into the atmosphere through the chimneys. They have a high temperature and steam having a high energy flows through the chimney without further use.

When gas is heated, large amounts of flue gas are generated and released into the atmosphere through the chimneys. They have high temperature and water vapor, which has high energy, escapes the chimney without any further use. With conventional gas boilers, the connection of the condensate hose to the waste is complicated. The hose may clog or freeze, and the condensate can enter the boiler and cause a boiler failure. Sewage water can also be pushed out into the boiler [28,29].

2.3. Application of the Methodology

A lot of methods (tools) have been utilized over the last decades to meet the different actions required in the conduction of impact studies. The objectives of the different actions vary, as do the usable methods for each. These methods are described for example in [30] or [31,32].

Guidelines of European Commission [33] provide information on methods and tools that were selected from case studies and literature research. These generally fall into two groups:

- Scoping and impact identification techniques—these identify how and where an indirect or cumulative impact or impact interaction would occur—Network and analysis; Consultation and questionnaires; Checklists, Matrices; Expert opinion.
- Evaluation techniques—these quantify and predict the magnitude and significance of impacts based on their context and intensity—Matrices; Expert opinion.

It is possible to apply at the outset to define the problem, establish the terms of reference, design the overall EIA process, and set the study boundary. Scoping helps to reform EIA institutional arrangements by focusing on each EIA activity and documents (impact identification and prediction, public and agency consultation and other), as well as subsequent proposal acceptance or rejection [34].

During the EIA process usually, a combination of techniques is used, or approaches are adopted at different stages of the project. Examples of both categories are set out below, in Figure 5.

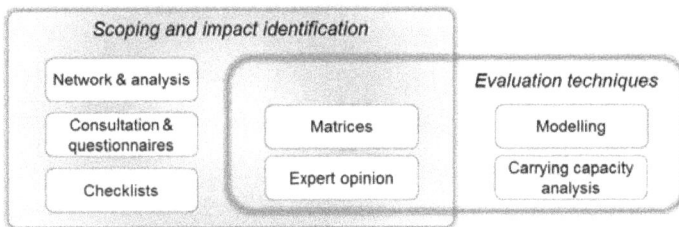

Figure 5. Methods and tools for assessment of indirect and cumulative impact as well as impact interaction (after Walker and Johnston 1999) [22].

The next differentiation of methods in the EIA process is given also in Zeleňáková and Zvijáková (2017) [22]. Table 1 presents the recommended methods suitable for impact assessment in the preliminary impact study, environmental impact statement and methods of multicriteria analysis.

Table 1. Methods for impact assessment in the preliminary impact study, environmental impact statement and methods of multicriteria analysis (after Kozová et al. 1996) [22].

Group of Methods	Methods
Recommended methods suitable for preliminary impact study	Ad hoc methods Checklists and catalogues of criteria Tables and matrices, expressing the causes and the effects Networks and system diagrams Decision trees Overlay mapping

Recommended methods suitable for environmental impact statement	Prognostic methods
	• Modeling
	  - mathematical models
	  - simulation models
	  - experimental in situ or laboratory models
	
	• Comparative methods

Multicriteria analysis	Method of utility function Total Environmental Quality Indicator (TIEQ) method Methods for determining weights of criteria • Ranking method • Allocation method • Grading method • Pairwise comparison method • Dual method ALO-FUL

Multicriteria analysis considers the assessment of the impact on the environment. Objective assessment is based on the use of quantitative factors—objective technical, technological, social and economic units. Qualitative assessment includes [35,36]: indicators; indices; intervals; categories; classes. Indicators can be represented by its description. Indices and intervals can express the magnitude or quality of parameters using the index function. Categories and classes use usually the subjective assessment of the features of interval and/or ratio scales. Formalized workflow involves ensuring that detection of an impact is done using a single method, and prognosis of induced changes is carried out on a scientific basis.

The selection of the optimal alternative is enjoyed by various methods, particularly by multicriteria analysis. The general procedure of multicriteria evaluation of alternatives includes six relatively discrete steps [8]:

• The creation of a purpose-oriented set of evaluation criteria;
• Setting the weights of the evaluation criteria;
• Assessment of the results (consequences, benefits, but also potential damages or losses of alternatives), it is a partial assessment of the alternatives;
• Assessment of the risks associated with implementing the alternatives;
• Determination of the preference order of alternatives and selection of the best option.

Multicriteria analysis is used to determine the value of a comprehensive land use in terms of the quality of the environment affected by humans. Multicriteria method utilizes the catalogue of criteria [8,37,38]. Its structure is hierarchical, adaptive and, basically, the whole society allows you to select the preferred option of a conventional set of alternatives or to give a preferential position of alternatives to a given set of criteria.

In the EIA process, it is always necessary to consider at least two alternatives to the proposed action:

(I) zero alternative—if there is no activity (current state of the environment); and
(II) alternatives of the proposed activity—variants of the activity that usually differ in the locality (site of construction), used technology, time of implementation, etc.

The purpose should be to find the optimal solution, in practice a choice called "preferred option". The selection of the optimal alternative is enjoyed by various methods, particularly by multicriteria analysis [36].

The evaluation presented in a previous paper [39] was done using the method of the Total Environmental Quality Indicator (TIEQ) to calculate the optimal variant. The TIEQ method is a modified utility function. The principle of this method is that the assessed impacts represented by the relevant indicators can be considered from the point of view of quantity and quality and transformed into partial utility functions and these values can already be compared [37]. According to this method alternative of biomass-fired power plant, (A1) is the best one [39]. However, due to the extent of the project, the demand for proof of this fact by another method was required. It makes the calculation more objective and has a higher value. The index coefficient method was chosen as the next method of multicriteria analysis for assessment and for selecting the best option for boiler construction in Trebišov city.

The index coefficient method determines the partial evaluation of the criteria through the calculation of the partial profitability through the index coefficient k_{ij}. The total profitability of the variant U_j (Equation (1)) is determined as the sum of the coefficients of the index coefficients k_{ij} (Equation (2)) and the criteria of the relevant variant and the weight of the *i*-th criterion [40]:

$$U_j = \sum P_{ij} = \sum k_{ij}.w_i \tag{1}$$

U_j is the overall profitability of the *j*-th variant; P_{ij} is the *i*-th criterion of the *j*-th variant; k_{ij} is the index coefficient of the *i*-th criterion of the *j*-th variant; w_i is the weight of the *i*-th criterion where $\Sigma w_i = 100$. The value of U_j determines the order of the advantage of the variants. There are also different possible methods for determining weights of criteria:

- Ranking method
- Allocation method
- Grading method
- Pairwise comparison method
- Dual method ALO-FUL

Pairwise comparison method by Fuller triangle was used in the presented case. The determined criteria were compared based on a number of preferences to each criterion. The total number of the pairs is n/2(n − 1); in our case, it is 9/2(9 − 1) = 36. The final weight for each criterion is calculated by a number of preferences to criterion divided by 36. The variation that achieves the highest value of overall performance is the best possible [40].

$$k_{ij} = \frac{H}{b_v} \tag{2}$$

The coefficient k_{ij} is determined by comparing the value of the criterion of the variant under consideration (H) with the value of the basic variant (b_v). A basic variant is a fictitious variant whose values of the criteria are theoretically the best (worst) or variants whose values for the criteria are predetermined according to the set goal. The principle of this method is that in a fictitious base variant we replace criterion values by number 1. The profitability of the criteria for the other variants is calculated using a coefficient as the ratio between the value of the calculated variance criteria and the original value of the base criterion. To determine the k_{ij} coefficient, two classes of criteria have to be distinguished. The first class is profit-type criteria, where higher values are preferred to lower performance levels (i.e., the higher the criterion value, the better). The second class is cost-type criteria where lower values are preferred to higher performance levels (i.e., the lower the criterion value, the better this criterion) [40]. Thus, criteria for increasing or decreasing preference are distinguished.

3. Results and Discussion

In the EIA process, considerations of two alternatives of the proposed action in comparison with zero alternative is usually taken in the assessment. In the presented case the study of considered alternatives are:

- Alternative 0—the zero alternative—if no activity is implemented.
- Alternative 1—the biomass-fired power plant in Trebišov district.
- Alternative 2—modernization of existing natural gas boiler.

Comparison of alternatives is done by multicriteria analysis. The beginning of this evaluation is creating a set of criteria and determining their importance (weight). We have chosen nine criteria described in Table 2. For evaluation and comparison, the criteria are divided into qualitative and quantitative ones.

Table 2. Description of criteria.

Aspect	Economic	Technical	Ecological	Social
Criteria (P_i)	the total cost of construction (P1)	time of construction (P3)	waste production (P6)	job opportunities (P8)
	annual operation cost (P2)	land occupation (P4)	emissions production (P7)	extra boiler room construction (P9)
		energy outputs of the power plants (P5)		
Character of assessment	quantitative	quantitative	qualitative	qualitative

The points (0–10) associated with each criterion (Table 3) were stated based on different experts' suggestions [23] (10 experts were involved) by brainstorming method with the aim to get the most objective results. The opinions of experts did not differ widely in most cases.

Table 3. Index coefficient method.

Criteria Pi/ Alternative Ai		P1	P2	P3	P4	P5	P6	P7	P8	P9	$\sum U_j$
b_v		9	8	10	8	9	8	7	9	8	
w_i		3	3	8	3	11	17	19	17	19	100
A0	value	0 €	0 €	0 months	0 m²	0 MW	no	0%	0	yes	
	points	9	8	10	8	0	8	7	0	0	
	k_{i0}	1	1	1	1	0	1	1	0	0	
	$k_{i0}.w_i$	3	3	8	3	0	17	19	0	0	53
A1	value	3.8 mil. €	0.54 mil. €	9 months	2000 m²	14 MW	yes	0%	8	no	
	points	3	6	2	3	9	5	7	9	8	
	k_{i1}	0.33	0.75	0.2	0.375	1	0.625	1	1	1	
	$k_{i1}.w_i$	0.99	2.25	1.6	1.125	11	10.63	19	17	19	82.6
A2	value	2.6 mil. €	0.65 mil. €	3 months	1250 m²	10 MW	yes	6.5%	6	no	
	points	8	5	8	6	8	4	6	7	8	
	k_{i2}	0.88	0.625	0.8	0.75	0.88	0.5	0.86	0.78	1	
	$k_{i2}.w_i$	2.64	1.875	6.4	2.25	9.68	8.5	16.3	13.26	19	79.9

Ecological criteria are the most important from an environmental perspective. The evaluation is influenced by subjective opinions of evaluators. The effects of operating variants cannot be measured, assessed and compared. Two criteria are evaluated (P6 and P7). A0 has the best results (no waste production) and A1 and A2 have broadly similar points (some waste production exists). During the combustion processes of biomass, the ash is produced and, in the case of modernized natural gas boiler, condensation and latent heat appear. A0 and A1 produce no emission, which is why they are better than A2.

Economic criteria are important aspects of comparison. In this case, they determine and compare the costs spent on the construction (P1) and annual operation of the alternatives (P2). In our case, the cheapest variant is Alternative A0, while Alternative 1 (A1) is more expensive in the total costs of the construction compared to Alternative 2 (A2), but is more favourable in the annual operating costs.

Technical and technological criteria were evaluated quantitatively. Criterion as the time of construction (P3), land occupation (P4) and energy output (P5) of three boilers belongs here. In two criteria, A0 is more advantageous than A1 and A2, but in the comparison with the third criterion—the output of boilers—Alternative 0 is the worst. Comparing criteria of A1 and A2 at the time of construction and land occupation, A2 is the preferred option, but according to an aspect of energy output, A1 seems to be a better variant.

Social criteria are also qualitatively assessed. The rating is influenced by subjective opinions of evaluators as well. Social impacts directly interfere with the quality of the life of the population, in our case, the alternative which offers more job opportunities (P8) and where it is needed for extra boiler room construction (P9). Since boilers are designed for citizens, they are designed to improve the quality of their lives. In our case, it has been compared which alternative can provide more job opportunities and which alternative contains the residents' need to build their own home boiler room. In both alternatives mentioned above, A0 is the worst. A1 is better than A2 compared to the job opportunities. In terms of the second criterion, both alternatives are able to compensate for the home boiler to a sufficient extent so that residents do not have to build the home boiler rooms.

The second step was to determine the index coefficient k_{ij}. This is determined by evaluating criteria where the value of the highest criterion is replaced by a value of 1. Subsequently, the other values are calculated as the ratio of value 1 to the appropriate variant value. The basic variant in this step is those where number 1 is located. The total profitability was calculated based on Equation (1).

The construction of the biomass-fired power plant is the best of the three compared alternatives based on the selected criteria from the point of view of the environmental impact assessment (Figure 6).

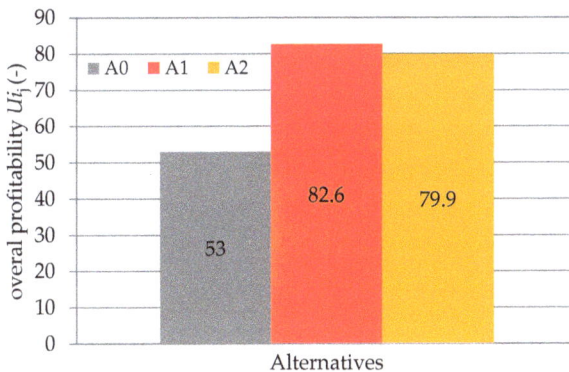

Figure 6. Summary comparison of the assessed alternatives.

The aim of the assessment was to make an overall investigation, description and evaluation of the direct and indirect environmental impacts of the activity on the environment, to determine measures that will prevent or mitigate pollution and damage to the environment, and to explain and compare the advantages and disadvantages of the proposed activity including its variants, in comparison also with the situation that would exist if the activity was not implemented. Biomass-fired power plant seems to be the better alternative (U_2 = 82.6) compared to modernizing the existing natural gas boiler (U_3 = 79.9) and compared to if no activity were implemented (U_0 = 53).

Alternatives 1 and 2 only have 2.7 points of difference. This difference is very low, so we made a sensitivity analysis of the different criteria, which are the criteria that have the real importance in the final decision. Sensitivity analysis of Alternative 0 is presented in Figure 7, Alternative 1 is presented in Figure 8 and Alternative 2 is presented in Figure 9 where criteria: P1—the total cost of construction, P2—annual operation cost, P3—the time of construction, P4—land occupation, P5—energy output, P6—waste production, P7—emissions production, P8—jobs opportunities, P9—extra boiler room construction.

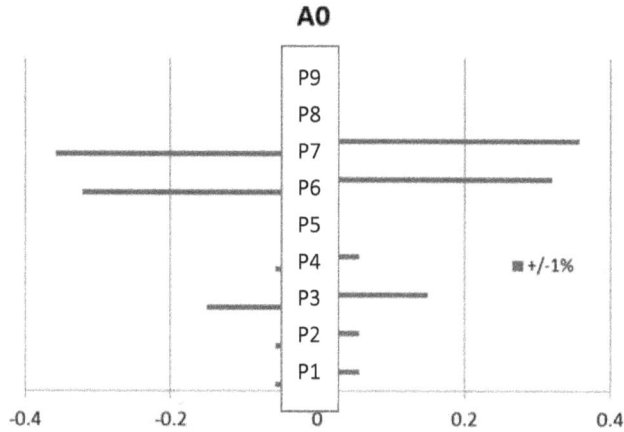

Figure 7. Sensitivity analysis of Alternative 0.

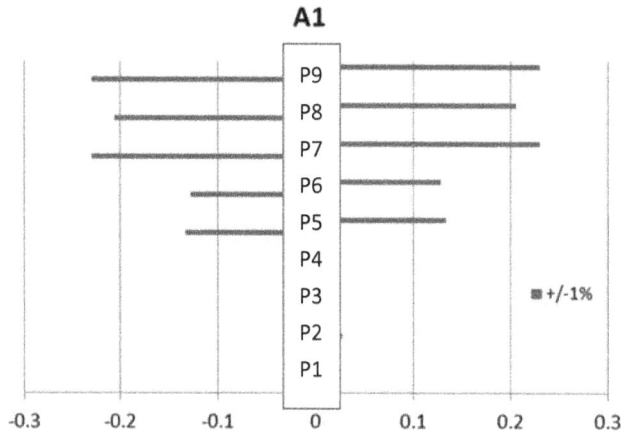

Figure 8. Sensitivity analysis of Alternative 1.

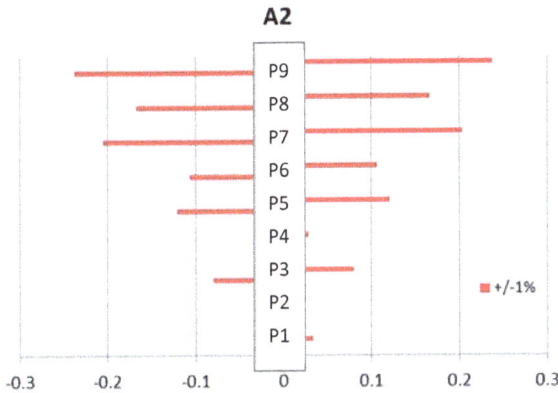

Figure 9. Sensitivity analysis of Alternative 2.

Figures 7–9 show results from sensitivity analysis as to how the model is sensitive to the choice of the criteria. The figures depict the effect on the selected activity (its alternatives) due to the different input data. Criteria: P7—emissions production and P9—extra boiler room construction have major influences on the selection of the best alternative. The assessment shows the highest sensitivity to these criteria. The proposed activity shows little sensitivity to all other criteria P1–P6 and P8.

4. Conclusions

The environmental impact assessment process for plants, structures, facilities and other activities has been applied in developed countries for several decades and is one of the main tools of preventive environmental protection and sustainable development.

The use of biomass in heating systems is beneficial because it uses agricultural, forest, urban and industrial residues and waste to produce heat and electricity with less effect on the environment than fossil fuels.

Coefficient index method as a tool of multicriteria analysis was used for proving that the construction of biomass-fired power plant is the most suitable solution chosen from three assessed variants (no activity is implemented, biomass power plant and modernization of existing gas boiler).

The index coefficient was used to state the weights of criteria. The points (0–10) associated with each criterion were stated based on ten different experts' suggestions with the aim to get the most objective results. In this evaluation, the highest score is the best possible. According to these points, Alternative A2 seems to be the best one (overall profitability of 82.6). Proposals were discussed with professionally qualified persons working in the field of environmental impact assessment as well as civil engineers.

The main contribution of the present paper is using theoretical knowledge of the issue, evaluation on the state of the environment in the area by the multicriteria analyses to select the optimal alternative of the action in the decision-making process in order to preserve environmental quality for further sustainable development of society in the study area. Most important for the assessment is the right selection of the criteria. In our assessment, we found out by a sensitivity analysis that the criteria that have the real importance in the final decision are P7—emissions production and P9—extra boiler room construction.

Author Contributions: Vlasta Ondrejka Harbulakova prepared and edited the manuscript; Martina Zelenakova worked on the materials and methods section—application of the methodology; Pavol Purcz worked on the sensitivity analysis; Adrian Olejnik prepared the expert part—description of the project and the study area.

Conflicts of Interest: The authors declare no conflict of interest.

References

1. Slávka, Štroffeková. Energy and Its Environmental Impact Assessment in the Slovak Republic in 2011. The Indicator Sectoral Report. Available online: https://www.enviroportal.sk/uploads/spravy/sprava-ener-2013-final.pdf (accessed on 25 January 2018). (In Slovak)
2. European Commision. Commission Staff Working Paper on Sustainable Industrial Development. Brussels, 25.10.1999, SEC(1999) 1729. Available online: http://ec.europa.eu/environment/archives/action-programme/pdf/sec991729_en.pdf (accessed on 9 December 2017).
3. Directive 2001/77/EC of the European Parliament and of the Council of 27 September 2001 on the Promotion of Electricity Produced from Renewable Energy Sources in the Internal Electricity Market. Available online: https://publications.europa.eu/en/publication-detail/-/publication/9df8db5d-4776-4fcb-9077-4363e14b836a (accessed on 9 December 2017).
4. Directive 2003/54/EC of the European Parliament and of the Council of 26 June 2003 Concerning Common Rules for the Internal Market in Electricity and Repealing Directive 96/92/EC—Statements Made with Regard to Decommissioning and Waste Management Activities. Available online: https://publications.europa.eu/en/publication-detail/-/publication/caeb5f68-61fd-4ea8-b3b5-00e692b1013c (accessed on 9 December 2017).
5. Impact of Electricity Production on the Environment. Available online: https://online.sse.sk/portal/page/portal/stranka_SSE/spravy/ekologia (accessed on 30 November 2017). (In Slovak)
6. Commission of the European Communities. Commission Staff Working Document. Accompanying Document to the Communication from the Commission to the Council and the European Parliament, Renewable Energy Road Map Renewable Energies in the 21st Century: Building a More Sustainable Future, Impact Assessment. Brussels, SEC (2006) 1719/2. Available online: http://www.ebb-eu.org/legis/renewable%20energy%20roadmap%20full%20impact%20assessment%20100107.pdf (accessed on 9 December 2017).
7. Directive 2009/28/EC of the European Parliament and of the Council of 23 April 2009 on the Promotion of the Use of Energy from Renewable Sources and Amending and Subsequently Repealing Directives 2001/77/EC and 2003/30/EC (Text with EEA Relevance). Available online: http://eur-lex.europa.eu/eli/dir/2009/28/oj (accessed on 9 December 2017).
8. Zvijaková, L.; Zeleňáková, M. *Risk Analysis in Environmental Impact Assessment of Flood Protection Measures*; Leges: Prague, Czech Republic, 2015.
9. Kumar, S.; Katoria, D.; Sehgal, D. Environment Impact Assessment of Thermal Power Plant for Sustainable Development. *Int. J. Environ. Eng. Manag.* **2013**, *4*, 567–572.
10. NTPC Limited. *Environmental Impact Assessment Report for Tanda Thermal Power Project, Stage–II (2 × 660 mw)*; District-Ambedkar Nagar (UP), Document No.: 9562/999/GEG/S/001 Rev. No. 1; NTPC Limited: New Delhi, India, 2009.
11. Pokale, W.K. Effects of thermal power plant on environment. *Sci. Rev. Chem. Commun.* **2012**, *2*, 212–215.
12. Cigognetti, G.; Piccardim, M.; Vital, C. Ministero dell Ambiente e Della Tutela del Territorio e del Mare, Thermal Power Plant of Ponti sul Mincio—Conversion of the Chimney into Lookout Tower and Mincio Park Entrance as a Change to the Prescription of the EIA Screening Determination 2014. Available online: http://www.va.minambiente.it/en-GB/Oggetti/MetadatoDocumento/109647 (accessed on 3 January 2018).
13. Coal Power Generation Comp any of Bangladesh Limited (An Enterprise of the People's Republic of Bangladesh). Report on Environmental Impact Assessment of Construction of Matarbari 600X2 MW. 2013. Available online: https://libportal.jica.go.jp/library/Data/DocforEnvironment/EIA-EPC/EastAsia-SouthwesternAsian/ChittagongACFPPDP/BCEIA.pdf (accessed on 3 January 2018).
14. Bo, X.; Wang, G.; Meng, F.; Wen, R. Air pollution effect of the thermal power plants in Beijing-Tianjin-Hebei region. *China Environ. Sci.* **2015**, *35*, 364–373.
15. Sun, Y.; Liu, Y.; Liu, C. Environmental impact assessment on ambient air fine particulate matter $PM_{2.5}$ role of prevention. *J. Shenyang Jianzhu Univ. (Nat. Sci.)* **2013**, *29*, 1147–1152.
16. Enríquez-de-Salamanca, Á.; Martín-Aranda, R.M.; Díaz-Sierra, R. Consideration of climate change on environmental impact assessment in Spain. *Environ. Impact Assess. Rev.* **2016**, *57*, 31–39.

17. Say, N.P.; Yücel, M.; Yilmazer, M. A computer-based system for environmental impact assessment (EIA) applications to energy power stations in Turkey: ÇEDINFO. *Energy Policy* **2007**, *35*, 6395–6401. [CrossRef]
18. Moridi, F.; Atabi, F.; Nouri, J. Air Pollution Management based on the Selection of Appropriate Technologies for Air Pollutants Filtration using Multi-Criteria Decision-Making. *An. Acad. Bras. Cienc.* **2015**, *87*, 314–323.
19. Shah, A.; Salimullah, K.; Sha, M.H.; Razaulkah, K.; Jan, I.F. Environmental impact assessment (EIA) of infrastructure development projects in developing countries. *OIDA Int. J. Sustain. Dev.* **2010**, *1*, 47–54.
20. Directive 2014/52/EU of the European Parliament and of the Council of 16 April 2014 on the Assessment of the Effects of Certain Public and Private Projects on the Environment. 2014. Available online: http://ec.europa.eu/environment/eia/pdf/EIA_Directive_informal.pdf (accessed on 25 January 2018).
21. National Council of the Slovak Republic 2005 Act of Law No. 24/2006 from December 14th 2005 on Environmental Impact Assessment and on Amendments to Certain Acts. Available online: http://www.ujd.gov.sk/files/legislativa/145_2010_EN.pdf (accessed on 25 January 2018).
22. Zeleňáková, M.; Zvijáková, L. *Using Risk Analysis for Flood Protection Assessment*; Springer: Cham, Switzerlandm, 2017; p. 128.
23. Trebisovksa Energeticka Inc. *Central Energy Source of Trebisov*; TM-P-114.13-B, Technical Report—Summary; Trebisovksa energeticka Inc.: Trebisov, Slovakia, 2013. (In Slovak)
24. Olejník, A. Environmental Impact Assessment of Construction. Master's Thesis, Technical University of Kosice, Košice, Slovakia, 6 May 2016.
25. Maphil.com, Free Satellite Location Map of Rad. Available online: http://www.maphill.com/slovakia/kosice/trebisov/location-maps/satellite-map/ (accessed on 30 November 2017).
26. Slámková, M.; Garčárová, M. Encouraging the Protection of Natura 2000 Sites in Integrating Whole-System of Ecological Stability. Regional Territorial System of Ecological Stability of the Trebišov District. 2012. Available online: www.minv.sk/?verejne-vyhlasky-7&subor=207452 (accessed on 30 November 2017). (In Slovak)
27. Ivaň, M. Mechanisms for Recovery of Wastes of Category "O" Others, Recovery of Construction Debris, Paper, Plastics, Wood," Company Trebišov Town, LAND—Service 2011. Enviroportal.sk, Information Portal of MoE SR. Available online: http://www.enviroportal.sk/sk_SK/eia/detail/ing-michal-ivan-land-servis-trebisov-zariadenie-na-zhodnocovanie-odpad (accessed on 30 November 2017). (In Slovak)
28. Banik and Son Inc. Differences between Traditional and Condensing Boilers. 2015. Available online: http://www.banik.sk/rozdiely-medzi-klasickym-a-kondenzacnym-kotlom/ (accessed on 30 November 2017). (In Slovak)
29. Durikova, K. How Does Condensing Boiler Works? Bratislava: JAGA Group. 2013. Available online: http://mojdom.zoznam.sk/cl/10055/1351440/Oplati-sa-investovat-do-plynoveho-kondenzacneho-kotla- (accessed on 30 November 2017). (In Slovak)
30. Canter, L.W. Methods for Effective Environmental Information Assessment (EIA) Practice. In *Environmental Methods Review: Retooling Impact Assessment for the New Century*; Porter, A.L., Fittipaldi, J.J., Eds.; The Press Club: Fargo, ND, USA, 1998.
31. Canter, L.W. Environmental Impact Assessment. In *Environmental Engineers' Handbook*; Liu, D.H.F., Lipták, B.G., Eds.; CRC Press: Boca Raton, FL, USA, 1999.
32. Kozova, M.; Drdos, J.; Pavlickova, K.; Uradnicek, S. *Environmental Impact Assessment*; Comenius University in Bratislava: Bratislava, Slovakia, 1996.
33. Walker, L.J.; Johnston, J. Guidelines for the Assessment of Indirect and Cumulative Impacts as well as Impact Interactions. 1999. Available online: http://ec.europa.eu/environment/archives/eia/eia-studies-and-reports/pdf/guidel.pdf (accessed on 25 January 2018).
34. Lawrence, D.P. *Environmental Impact Assessment: Practical Solutions to Recurrent Problems*, 2nd ed.; John Wiley & Sons: Hoboken, NJ, USA, 2013.
35. Říha, J. *Environmental Impact Assessment. Methods for Preliminary Decision Analysis*; ČVUT: Praha, Czech Republic, 2001. (In Czech)
36. Říha, J. *Environmental Impact Assessment of Investments*; Multicriteria Analysis and EIA; Academia: Prague, Czech Republic, 1995.
37. Majerník, M.; Húsková, V.; Bosák, M.; Chovancová, J. *Methodology of the Environmental Impact Assessment*; Technical University of Košice (TUKE): Košice, Slovakia, 2008.

38. Gałaś, S.; Krol, E. Indicators for environmental-spatial order assessment on the example of the Busko and Solec Spa communes. *Gospodarka Surowcami Miner.-Miner. Resour. Manag.* **2008**, *24*, 95–115.
39. Zelenakova, M.; Ondrejka Harbulakova, V.; Olejnik, A. Using of Multicriteria Method for Choosing the Best Alternative of the Heating Power Plant, Selected Scientific Papers. *J. Civ. Eng.* **2017**, *12*, 47–56.
40. Kampf, R. Estimation Methods for Weight Criteria. *Sci. Pap. Univ. Pardubice B* **2003**, *9*, 255–261.

environments

MDPI

Case Report

Overview of Green Building Material (GBM) Policies and Guidelines with Relevance to Indoor Air Quality Management in Taiwan

Wen-Tien Tsai

Graduate Institute of Bioresources, National Pingtung University of Science and Technology, Pingtung 912, Taiwan; wttsai@mail.npust.edu.tw; Tel.: +886-8-770-3202; Fax: +886-8-774-0134

Received: 30 October 2017; Accepted: 13 December 2017; Published: 28 December 2017

Abstract: The objective of this paper was to offer a preliminary overview of Taiwan's success in green building material (GBM) efforts through legal systems and promotion measures, which are relevant to the contribution to indoor air quality (IAQ) due to sustainability and health issues. In the first part of the paper, the IAQ regulations are summarized to highlight the second nation (i.e., Taiwan) around the world in IAQ management by the law. In addition, the permissible exposure limits (PEL) in Taiwan for airborne hazardous substances were first promulgated in 1974 to deal with occupational health issues in the workplace environment. In the second part of the paper, the developing status of the GBM in Taiwan is analyzed to unravel its connection with the Indoor Air Quality Management Act (IAQMA), promulgated on 23 November 2011. By the end of September 2017, a total of 645 GBM labels have been conferred, covering over 5000 green products. Due to the effectiveness of source control, the healthy GBM occupies most of the market, accounting for about 75%. The IAQMA, which took force in November 2012, is expected to significantly increase the use of healthy GBM in new building construction and remodeling, especially in low formaldehyde (HCHO)/volatile organic compound (VOC)-emitting products.

Keywords: green building material; indoor air quality; occupational exposure; volatile organic compound; source control

1. Introduction

Most people spend over two thirds of their time indoors. The indoor air at home or in an office building, school and other workplaces could be contaminated by a variety of gaseous and particulate contaminants that are sometimes present in concentrations above those which cause adverse health effects. These indoor air pollutants (IAP) are mainly emitted from building materials, furnishings, office appliances/equipment, consumer products, cleaning/maintenance materials, combustion processes (e.g., tobacco smoking, fuel-fired cooking or space heating), and outdoor air pollution. Regarding the regulations for ambient air quality, for instance, the Clean Air Act Amendments of 1990 in the USA was enacted and revised to make the law more readily enforceable. Based on the field investigations, the concentrations of individual volatile organic compounds (VOC) in the indoor air are often higher than those outdoors because many building materials emit VOCs over their extended periods of time [1]. As a consequence, the indoor air quality (IAQ) has become an important health issue for the public and the decision makers because of its adverse impact on acute or chronic symptoms or illness [2,3]. The sick building syndrome (SBS) has been used to describe building-related symptoms [4], which include respiratory irritation, headache, dry cough, dry or itchy skin, dizziness and nausea, difficulty in concentrating, fatigue, and sensitivity to odors [5]. In addition, formaldehyde (HCHO), a potent mucous membrane irritant and a widespread IAP, has been listed as a Class 1 carcinogen (confirmed as a human carcinogen) by the International Agency for Research on Cancer [6].

Therefore, many national/federal competent authorities (e.g., the Environmental Protection Agency, the Consumer Product Safety Commission, the Department of Housing and Urban Development, and the Occupational Safety and Health Administration in the United States) have stipulated legal standards and/or voluntary guidelines that involved in indoor/workplace air quality management, and dangerous/defective products [7].

In response to the global trends in climate change mitigation and IAQ-based health issues since the 1980s, the industry, officials and universities have jointly committed to promote the green building (GB) or sustainable building concept [8]. Thereafter, the GB certification system was developed in the United States and European countries [9]. To achieve sustainability and resolve health issues, GB schemes usually include several prevention and control measures for building design and construction. Among them, the adoption of certified healthy green building material (GBM) is an efficient scheme because the sources of IAP come originally from a variety of building materials. With the use of low-emission products, the problem buildings, especially in new building construction and remodeling, can be significantly reduced [10]. For instance, alternative products with low levels of HCHO can be used instead of HCHO-emitting products such as particleboard and hardwood plywood paneling. As described later [11], based on the regulatory definition in Taiwan, healthy GBMs have better IAQ than conventional building materials, or so-called green, natural or organic products or materials because the latter cases can contain toxic or hazardous constituents [12].

Through the media's reports on the monitoring results of IAQ by academic scholars and non-profit organizations, Taiwanese people have been increasingly paying attention to the adverse impact of IAPs on human health since the early 2000s. This has led to the necessity of legislation for the purposes of protecting public health and also improving work performance. The Indoor Air Quality Management Act (IAQMA) was promulgated on 23 November 2011, and took effect one year after promulgation. In the Act, the central competent authority refers to the Environmental Protection Administration (EPA). Also, the IAPs refers to substances that are normally dispersed in indoor air, and which may directly or indirectly affect public health or the living environment after long term exposure, including carbon dioxide, carbon monoxide, formaldehyde, total volatile organic compounds (TVOC), bacteria, fungi, airborne particles ($PM_{2.5}$ and PM_{10}), ozone, and other substances designated and officially announced by the EPA. On the other hand, the Ministry of Labor (MOL) has committed to the Occupational Safety and Health Administration (OSHA) to deal with IAQ-related issues in the workplace environment. Therefore, the permissible exposure limits (PEL) in Taiwan for airborne hazardous substances in workplace were promulgated since the 1970s, and recently revised in 2014. Meanwhile, the Architecture and Building Research Institute (ABRI) under the Ministry of Interior (MOI) drafted the GBM labeling system in 2003, and formally implemented it in 2004. The core value of the GBM is based on the non-toxicity, harmlessness, and relevant specification standards met. Currently, there are four GBM categories, covering the features of ecology, health, recycling, and high-performance.

This paper offers a preliminary analysis of the IAQ improvement through of the joint efforts of the cross-ministries in Taiwan, including the EPA, the MOL and the MOI. In the first part of the paper, the IAQ regulations are summarized to highlight the features of the IAQMA because Taiwan became the second nation when South Korea introduced their regulations dedicated to IAQ control. In line with the international trends in GB and/or sustainable building in recent years, the developing status of the GBM in Taiwan is analyzed in the second part of the paper to unravel its connection with the IAQ.

2. Indoor Air Quality Regulations in Taiwan

2.1. Environmental Protection Administration

In order to define the IAQ within a built environment, the IAQ guidelines and/or standards have been developed by several national and international agencies [7,13]. In some countries (e.g., South Korea), the IAQ standards are adopted by the regulatory authorities as an enforceable act or law. By contrast, the IAQ guidelines are designed by several countries or international organizations to offer only guidance to reduce adverse health impacts resulting from IAP in the living environment. For example, the World Health Organization (WHO) first published the IAQ guidelines for Europe in 1987, which contained health risk assessments of 28 air contaminants (especially formaldehyde and VOCs) [14]. The guidelines aim to provide a scientific basis for public health professionals, as well as specialists and authorities. However, it should be noted that the IAQ guidelines and/or standards for VOCs (except for formaldehyde) are generally described as the total volatile organic compounds (TVOCs).

In the past decades, it has been reported that the IAQ in Taiwan was not good [11,15], causing a large number of problem buildings. Their poor performance may be attributed to the crowed living space, poor air circulation, lack of ventilation, building materials, and humid subtropical climate. This problem has attracted much attention because most people spend about 90% of their time indoors, thus exposing to a variety of IAP that may be harmful to human health. In 2005, the EPA announced its "Suggested Values for Indoor Air Quality" and began research and preparations for a draft of the IAQMA in the following year. Through the public hearing and legislative procedures, the act aims to improve IAQ and to protect public health; it was promulgated on 23 November 2011, and came into effect one year later. Thereafter, the EPA announced five new regulations to accompany the implementation of the IAQMA. They include the "Indoor Air Quality Act Enforcement Rules", the "Indoor Air Quality Standards", the "Regulations Governing Dedicated Indoor Air Quality Management Personnel", the "Regulations Governing Indoor Air Quality Analysis Management", and the "Fine Determination Criteria for Violations of the Indoor Air Quality Act". In Taiwan, the IAQ standards, as listed in Table 1, provide compulsory guidelines. It shows 1000 ppm for CO_2, 9 ppm for CO, 0.06 ppm for formaldehyde, 0.56 ppm for TVOC (a combination of 12 different VOCs), and 0.06 ppm for ozone. Figure 1 shows a schematic block chart regarding historical development of IAQ codes and standards in Taiwan.

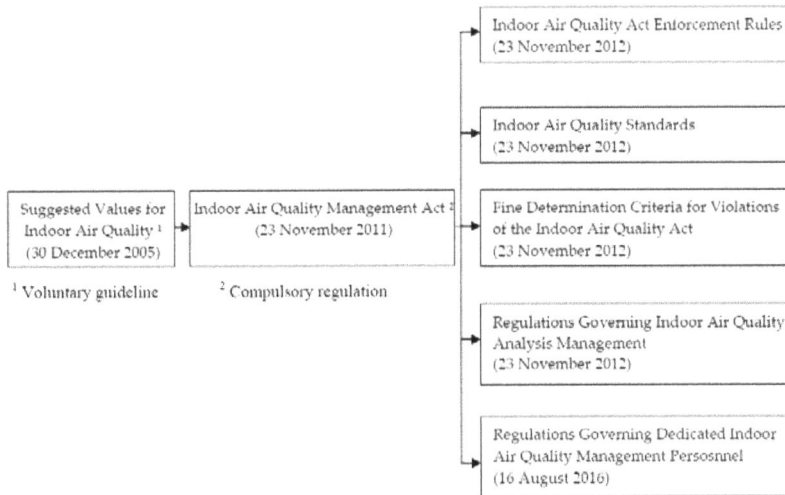

Figure 1. Indoor air quality (IAQ) regulations in Taiwan, providing the authority for the Environmental Protection Administration (EPA).

To facilitate the implementation of the IAQMA, the EPA announced the premises that shall comply with the Act, including libraries, hospitals and clinics, social welfare premises, government offices, railway/civil aviation/subway mail halls, exhibition centers, shopping malls, and so on. The owners, managers or users of officially announced premises shall commission certified analysis agencies to regularly perform analyses of IAQ based on the standards in Table 1, and shall regularly make public the analysis results and make records. On the other hand, the premises that have been officially announced by the EPA shall install automatic monitoring systems for continuous monitoring of IAQ, immediately make public the latest automatic monitoring results inside the premises or display them in a prominent place at the entrance, and shall make records.

Table 1. Indoor air quality standards as compulsory guidelines in Taiwan.

Indoor Air Pollutants	Standard	
	Concentration	Sampling Time
Carbon dioxide (CO_2)	1000 ppm	8 h
Carbon monoxide (CO)	9 ppm	8 h
Formaldehyde (HCHO)	0.06 ppm	1 h
TVOC [a]	0.56 ppm	1 h
Bacterial	1500 CFU/m^3 [b]	Ceiling
Fungi	1000 CFU/m^3 [c]	Ceiling
Particulate matter (PM_{10})	75 µg/m^3	24 h
Particulate matter ($PM_{2.5}$)	35 µg/m^3	24 h
Ozone (O_3)	0.06 ppm	8 h

[a] 12 volatile organic compounds, including benzene, carbon tetrachloride, chloroform, 1,2-dichlorobenzene, 1,4-dichlorobenzene, dichloromethane, ethyl benzene, styrene, tetrachloroethylene, toluene, trichloroethylene, and xylenes. [b] Colony-forming unit (CFU). [c] Not limited to the ratio of outdoor fungi concentration to indoor fungi concentration less than or equal to 1.3.

2.2. Ministry of Labor (MOL)

As described above, inhalation and skin (e.g., Beko et al. [16]) are the main routes of exposure to VOCs and other air contaminants in indoor environments, including workplace environment and living environment. Therefore, a healthy indoor environment plays a vital role in the prevention of damage to public health from health risks because these IAP are known or potential human carcinogens, or may be reasonably anticipated to pose a threat of adverse human health effects, including respiratory illness and neurological symptoms. Therefore, many national competent authorities (e.g., the Occupational Safety and Health Administration in the United States) and non-profit scientific organizations (e.g., the American Conference of Governmental Industrial Hygienists) have stipulated the standards or guideline levels which aim at limiting exposure to certain hazardous air contaminants in the workplace environment. Regarding the occupational exposure limit (OEL): this is defined as an acceptable concentration of a hazardous substance or class of hazardous materials in workplace air for the purpose of protecting the health of workers during their work. Among them, the most common OELs are the permissible exposure limits (PELs) and the threshold limit values (TLVs), which are developed by the U.S. Occupational Safety and Health Administration (OSHA) and the American Conference of Governmental Industrial Hygienists (ACGIH), respectively.

In Taiwan, the OEL ("Standards of Permissible Exposure Limits of Airborne Hazardous Substances in Workplace") was promulgated in August 1974 and was recently revised in June 2013 by the Ministry of Labor (MOL) under the authorization of the Occupational Safety and Health Act (OSHA) [17]. According to the definition by the OSHA, the workplaces referred to places where work for specific purposes takes place. However, the OEL standards in Taiwan apply only for industrial workplaces, plants, or factories. Initially, most of the PEL values in Taiwan were directly adopted from the data on the ACGIH-TLV and the OSHA-PEL. Thereafter, the Institute of Labor, Occupational Safety and Health (ILOSH), a national research & development organization under the supervision of the MOL, organized the Committee to recommend regularly the revised OEL for candidate chemicals based on the updated toxicological and epidemiological literature, policy need, and economical & technical feasibility. Currently, 487 chemicals have been listed in the PEL values as compulsory guidelines, including the time weighted average for an eight-hour workday, the time weighted average for short term exposure, and the ceiling. Because the TLV values are mainly based on health factors, nearly all workers may be repeatedly exposed to conditions below the PEL values without adverse effects. According to the IAQ standards in Table 1, Table 2 further lists the corresponding PEL and TLV values for these designated IAP. Table 2 also summarizes their TLV basis, representing the adverse health effects upon which the TLV is based [18]. In brief, the target organs or systems for human exposure to these designated IAP in the workplace air include respiratory tract, central nervous system (CNS), kidney and liver.

Table 2. Occupational exposure limits for designated air pollutants in indoor environment.

Contaminant	PEL (Taiwan) [a]	TLV (USA) [b]	
	Concentration (ppm)	Concentration (ppm)	Basis
Benzene	1	0.5	Leukemia
Carbon dioxide	5000	5000	Asphyxia
Carbon monoxide	35	25	Carboxyhemoglobinemia
Carbon tetrachloride	2	5	Liver damage
Chloroform	10	10	Liver & embryo/fetal damage; central nervous system (CNS) impairment
1,2-Dichlorobenzene	50	25	Upper respiratory tract (URT) & eye irritation; liver damage
1,4-Dichlorobenzene	75	10	Eye irritation; kidney damage
Dichloromethane	10	50	Carboxyhemoglobinemia; CNS impairment
Ethyl benzene	100	20	URT & eye irritation; kidney damage (nephropathy);cochlear impairment
Formaldehyde	1	0.3	URT & eye irritation
Ozone	0.1	0.05–0.2 [c]	Pulmonary function
Styrene	50	20	CNS impairment; URT irritation; peripheral nephropathy
Tetrachloroethylene	50	25	CNS impairment
Toluene	100	20	Visual impairment; female reproductive; pregnancy loss
Trichloroethylene	50	10	CNS impairment; cognitive decrements; renal toxicity
Xylenes	100	100	URT & eye irritation; CNS impairment

[a] Compulsory guidelines. [b] Source [18]. [c] Depending on work types (i.e., heavy, moderate, or light work). PEL: permissible exposure limits. TLV: threshold limit value.

3. Green Building Materials (GBM) in Taiwan

3.1. Governing Regulations

In order to achieve sustainable, comfortable and healthy living environments, the central competent authorities in Taiwan, including the Ministry of the Interior (MOI) and the Environmental Protection Administration, promulgated the regulations for promoting their recognized green-mark products [19]. In 2004, the Architecture and Building Research Institute (ABRI) under the MOI established and launched the Green Building Material Evaluation and Labeling System. These promulgated measures were originally based on the Basic Environment Act (BEA) passed by the Legislative Yuan (the Congress) at the end of 2002. For example, the BEA requires the government departments to adopt necessary measures to promote the use of renewable resources and other materials, products or services that beneficially lessen environmental impact. In addition, the Legislative Yuan passed the Government Procurement Law in 1997, in which it requires the official organizations to adopt environmentally preferable products certified by the EPA.

The GBM system aims at both taking into account a sustainable built environment and a healthy living quality. In order to attain the achievable goals, the MOI promulgated the Green Building Guidelines by several enforcement stages:
- Effective 1 July 2006

The percentage of GBM usage must be raised up to 30% for all decorating materials used in the interior furnishing and flooring of new and/or retrofitted buildings for public use.

- Effective 1 July 2012

The minimum requirement for the application ratio of GBM has to exceed 45% of the total indoor surface area for interior decoration materials, floor materials and windows, and 10% of outdoor surface area (deducting car lane, buffer space of car access, activity space of fire truck disaster relief, and part of laying ground material not needed) for the ground materials.

According to the definition by the Building Technique Regulation under the authorization of the Building Act, the constituents of the GBM shall meet one of the following requirements:

- Recycled plastic or rubber products (not containing toxic chemical materials designated by the EPA).
- Insulation materials for building use (not contain the substances controlled under the Montreal Protocol, and toxic chemical materials designated by the EPA).
- Water-based coatings/paints (not contain formaldehyde, chlorinated solvents, heavy metals such as mercury, lead, hexavalent chromium and arsenic; not use triphenyl tin and tributyl tin).
- Recycled wood products.
- Recycled bricks (kiln-burned) for building use.
- Recycled building materials (not kiln-burned).
- Other building materials certified by the central competent authorities (MOI or EPA).

Based on the above-mentioned regulations, the GBM must be non-hazardous to the environment, non-toxic to human health, and be in accordance with the national specifications/standards. Therefore, the evaluation items of the GBM include the following restricted substances:

- The concentrations of heavy metals involved in any part of non-metal materials must meet the "Toxicity Characteristic Leaching Procedure (TCLP) of Industrial Waste" set by the EPA (Table 3).
- Not contain asbestos.
- Not contain radioactive materials or constituents.
- Not contain the toxic chemical substances designated by the Toxic Chemical Substances Control Act (TCSCA). Under the designation of the act, there are currently 323 toxic chemical substances.
- Not contain the ozone-depleting substances controlled under the Montreal Protocol.
- Total chlorine ion content in the cement-based products must be less than or equal to 0.1%.
- Chlorine-containing polymers shall not apply for healthy GBM and ecological GBM labels.
- Interior decoration materials should be carried out the emission tests of total volatile organic compounds (TVOC) and formaldehyde by GBM performance testing agencies, which should be certified by the MOI. Their emission rates should at least meet the E3 rating of healthy GBM (Table 4), in which the rating standards are less than 0.05 and 0.19 mg/m^2·h for formaldehyde and TVOC, respectively.

Table 3. The limits of heavy metals involved in green building material (GBM).

Heavy Metal	Detection Standard (mg/L) [a]
Total mercury (T-Hg)	0.005
Total cadmium (T-Cd)	0.3
Total lead (T-Pb)	0.3
Total arsenic (T-As)	0.3
Hexavalent chromium (Cr $^{+6}$)	1.5
Total copper (T-Cu)	0.15
Total silver (T-Ag)	0.05

[a] According to "Toxicity Characteristic Leaching Procedure (TCLP) of Industrial Waste" set by the Environmental Protection Administration (EPA) in Taiwan.

Table 4. Rating system of healthy GBM in Taiwan.

Rating System	Emission Rate (mg/m²·h)	
	TVOC [a]	Formaldehyde (HCHO)
E1	≤ 0.005	≤ 0.005
E2	0.005 < TVOC ≤ 0.06	0.005 < HCHO ≤ 0.02
E3	0.06 < TVOC ≤ 0.19	0.02 < HCHO ≤ 0.05

[a] Total volatile organic compounds (TVOC) include benzene, carbon tetrachloride, chloroform, 1,2-dichlorobenzene, 1,4-dichlorobenzene, dichloromethane, ethyl benzene, styrene, tetrachloroethylene, trichloroethylene, toluene, and xylenes.

Currently, there are four GBM types, including:
- Ecological GBM

They refer to those which are made of renewably natural resources (e.g., woody materials) with easy biodegradation, low manual processing (low power input), and in conformity with the local industry (carbon footprint thus reduced) or the international certificates like the Forest Stewardship Council (FSC).
- Healthy GBM

They refer to those featuring low emissions of formaldehyde and total volatile organic compounds (TVOC). Based on their emission rates (Table 4), the rating system of healthy GBM further categorizes the E1, E2 and E3.
- High-performance GBM

They refer to those that display high performance of sound-insulation (noise prevention), energy-saving (heat insulation) or permeability (water drainage) without traditional defects of building materials/accessories.
- Recycled GBM

They refer to those which are remanufactured from local recycled materials to meet waste reduction, reuse and recycling (3R) requirements without causing secondary pollution or having a bad effect on human health.

Due to its acute/chronic toxicity and widespread uses in a variety of indoor products such as particleboard, fiberboard and hardwood plywood, many countries adopted IAQ guidelines or standards for formaldehyde (HCHO) exposures [12]. In addition, a large variety of synthetic organic compounds are found in indoor air because they are also emitted from a variety of commercial commodities, including underlayment, paneling, furniture, cabinetry, cleaning products and consumer products [1]. Because of their variety, relatively low concentrations and potential to cause sick building syndrome or building-related illness symptoms, the effect of the TVOC on human health have received much attention in connection with its IAQ guidelines or standards. As listed in Table 4, the rating system of healthy GBM in Taiwan is thus based on the emission rates of HCHO and TVOC, grouping building materials (including floor, wall, ceiling, gap-filler & putty, painting & coating, adhesive & bonding agent, and wooden door/window) into the E1, E2 and E3. Although TVOC as a health risk index or indicator was debatable [20], it is often used as a guide to determine whether VOC levels are elevated in indoor air, implying the potential for occupant irritation and discomfort [7].

3.2. Developing Status of GBM

The history of green building (GB) must be briefly described prior to the information about the developing status of GBM in Taiwan. The GB began in the late 1980s when the United Nations' World Commission on Environment and Development addressed the concept of "sustainability" in 1987 [21]. In the early 1990s, many developed countries developed the GB certification systems, including the Leadership in Energy and Environmental Design (LEED) certification in the United States and the BRE Environmental Assessment Method (BREEAM) certification in the United Kingdom [9].

In response to the international GB trend and the promotion of indoor environment quality, the central competent authority in Taiwan, (MOI) established the Green Building Evaluation System in 1998. Subsequently, the MOI launched the Green Building Labeling System in 1999, which was also called the EEWH labeling system. The GB labeling system involves the principles of ecology, energy-saving, waste-reducing and health. Currently, the system comprises of nine evaluation indicators, including greenery (vegetation planting), water infiltration and retention, daily energy conservation, water conservation, CO_2 emission reduction, construction waste reduction, sewage and waste disposal facility improvement, biodiversity, and indoor environment quality. On the other hand, the Executive Yuan ratified the Green Building Promotion Plan in 2001 to extend its implementation to the government entities at all levels and private sectors under the funding support of the MOI.

In order to accelerate the development of GBM, the ABRI, combining with the architecture department of national universities and non-profit organizations (e.g., Taiwan Architecture & Building Center), has been assisting the industry upgrading to create a green environmental-friendly subtropical Taiwan. According to the statistics surveyed by the MOI (Table 5), some significant points were further addressed as follows:

- Over the 6 years, the certified GBM number indicates an increase of 38.0%, increasing from 474 by the end of July 2011 to 645 by the end of September 2017. This change could be attributed to the official promotion and validation for the reduction of indoor HCHO concentration [22], and the cost-down of GBM in Taiwan. Currently, a total of 645 GBM Labels have been conferred, which cover over 5000 green products.
- Among these certified GBM, the percentage distributions indicate no significant change in recent years due to the market demands and/or consumer preferences. The healthy GBM occupied most of the market, accounting for about 75%. With the IAQMA promulgated in 2011, it is expected to significantly increase the use of the healthy GBM in the near future.
- Under the encouragement of government policy for procuring the domestic green-mark (environmentally preferable) products (including energy-saving products, and water-saving products), the certified high-performance and recycled GBM products indicates significant increases of 55% and 47%, respectively.

Table 5. Statistics on certified GBM in Taiwan.

Category	May 2011 [a]		September 2017 [c]	
	Certified Number	Percentage [b]	Certified Number	Percentage
Healthy GBM	364	76.8%	487	74.4%
High-performance GBM	71	15.0%	110	16.8%
Recycling GBM	38	8.0%	56	8.6%
Ecological GBM	1	0.2%	1	0.2%
Total	474	100.0%	654	100.0%

[a] Source [11]. [b] The percentage is based on the ratio of the number of certified products per category to total certified GBM products. [c] Surveyed by the author using the official database of the ABRI (http://www.abri.gov.tw/).

4. Conclusions

In this paper, the recent legislations on IAQ management and GBM in Taiwan were reviewed and coupled. The following conclusions can be drawn below:

- The Indoor Air Quality Management Act (IAQMA), promulgated on 23 November 2011, took effect one year after promulgation. Under the authorization of the IAQMA, the IAQ standards provide compulsory guidelines in the non-industrial sectors.
- The permissible exposure limits of airborne hazardous substances in the indoor workplace were recently revised in June 2013 under the authorization of the Occupational Safety and Health Act (OSHA), providing compulsory guidelines in the industrial sector.

- According to the voluntary guidelines by the Building Technique Regulation under the authorization of the Building Act, the Green Building Material (GBM) was established and launched since 2004. Currently, a total of 645 GBM Labels have been conferred, accounting for about 75% by the healthy GBM occupied in the market. With the IAQMA promulgated in 2011, it is expected to significantly increase the use of the healthy GBM in the near future.

People spent most of their lifetime in indoor environments. Thus, the indoor air quality has a significant impact on human health, because many VOCs and other air toxins exist indoors, at concentrations even exceeding those in outdoor air. Therefore, the regulatory and ventilation measures to reduce indoor emissions and exposure concentrations in the densely populated indoors (e.g., hospitals, hairdressing salons, or metro system), and also investigate the relationship between human health risk and long-term exposure to VOCs, should be imperative. On the other hand, we should develop criteria to classify all detectable VOCs emitted from all types of GBM.

Conflicts of Interest: The author declares no conflict of interest.

References

1. Godish, T.; Davis, W.T.; Fu, J.S. *Air Quality*, 5th ed.; CRC Press: Boca Raton, FL, USA, 2015.
2. Berglund, B.; Brunekreef, B.; Knoppe, H.; Lindvall, T.; Maroni, M.; Molhave, L. Effects of indoor air pollutant on human health. *Indoor Air* **1992**, *2*, 2–25. [CrossRef]
3. Jones, A.P. Indoor air quality and health. *Atmos. Environ.* **1999**, *33*, 4535–4564. [CrossRef]
4. Nakaoka, H.; Todaka, E.; Seto, H.; Saito, I.; Hanazato, M.; Watanabe, M.; Mori, C. Correlating the symptoms of sick-building syndrome to indoor VOCs concentration levels and odour. *Indoor Built Environ.* **2014**, *23*, 804–813. [CrossRef]
5. Al horr, Y.; Arif, M.; Katafygiotou, M.; Mazroei, A.; Kaushik, A.K.; Elsarrag, E. Impact of indoor environmental quality on occupant well-being and comfort: A review of the literature. *Int. J. Sustain. Built Environ.* **2016**, *5*, 1–11. [CrossRef]
6. Tsai, W.T. Toxic volatile organic compounds (VOCs) in the atmospheric environment: Regulatory aspects and monitoring in Japan and Korea. *Environments* **2016**, *3*, 23. [CrossRef]
7. Abdul-Wahab, S.A.; En, S.C.F.; Elkamel, A.; Ahmadi, L.; Yetilmezsoy, K. A review of standards and guidelines set by international bodies for the parameters of indoor air quality. *Atmos. Pollut. Res.* **2015**, *6*, 751–767. [CrossRef]
8. Ravindu, S.; Rameezdeen, R.; Zuo, J.; Zhou, Z.; Chandratilake, R. Indoor environment quality of green buildings: Case study of an LEED platinum certified factory in a warm humid tropical climate. *Build. Environ.* **2015**, *84*, 105–113. [CrossRef]
9. Wei, W.; Ramalho, O.; Mandin, C. Indoor air quality requirements in green building certifications. *Build. Environ.* **2015**, *92*, 10–19. [CrossRef]
10. Osawa, H.; Tajima, M. Ventilation strategies for each kind of building and statutory regulations. In *Chemical Sensitivity and Sick-Building Syndrome*; Yanagisawa, Y., Yoshino, H., Ishikawa, S., Miyata, M., Eds.; CRC Press: Boca Raton, FL, USA, 2017; pp. 79–95.
11. Hsieh, T.T.; Chiang, C.M.; Ho, M.C.; Lai, K.P. The application of green building materials to sustainable building for environmental protection in Taiwan. *Adv. Mater. Res.* **2012**, *343–344*, 267–272. [CrossRef]
12. Steinemann, A.; Wargocki, P.; Rismanchi, B. Ten questions concerning green materials and indoor air quality. *Build. Environ.* **2017**, *112*, 351–358. [CrossRef]
13. Olesen, B.W. International standards for the indoor environment. *Indoor Air* **2014**, *14*, 8–26. [CrossRef] [PubMed]
14. World Health Organization (WHO). WHO Guidelines for Indoor Air Quality: Selected Pollutants. Available online: http://www.euro.who.int/__data/assets/pdf_file/0009/128169/e94535.pdf (accessed on 4 October 2017).
15. Lu, C.Y.; Lin, J.M.; Chen, Y.Y.; Chen, Y.C. Building-related symptoms among office employees associated with indoor carbon dioxide and total volatile organic compounds. *Int. J. Environ. Res. Public Health* **2015**, *12*, 5833–5845. [CrossRef] [PubMed]

16. Bekö, G.; Callesen, M.; Weschler, C.J.; Toftum, J.; Langer, S.; Sigsgaard, T.; Høst, A.; Kold Jensen, T.; Clausen, G. Phthalate exposure through different pathways and allergic sensitization in preschool children with asthma, allergic rhinoconjunctivitis and atopic dermatitis. *Environ. Res.* **2015**, *137*, 432–439. [CrossRef] [PubMed]

17. Shih, T.S.; Wu, K.Y.; Chen, H.I.; Chang, C.P.; Chang, H.Y.; Huang, Y.S.; Liou, S.H. The development and regulation of occupational exposure limits in Taiwan. *Regul. Toxicol. Pharm.* **2006**, *46*, 142–148. [CrossRef] [PubMed]

18. American Conference of Governmental Industrial Hygienists (ACGIH). *2016 TLVs and BEIs: Based on the Documentation of the Threshold Llimit Values for Chemical Substances and Physical Agent*; ACGIH: Cincinnati, OH, USA, 2016.

19. Tsai, W.T. Green public procurement and green-mark products strategies for mitigating greenhouse gas emissions- experience from Taiwan. *Mitig. Adapt. Strateg. Glob. Change* **2017**, *22*, 729–742. [CrossRef]

20. Andersson, K.; Bakke, J.V.; Bjorseth, O.; Bornehag, C.G.; Clausen, G.; Hongslo, J.K.; Kjaellman, M.; Kjrgaard, S.; Levy, F.; Molhave, L.; et al. TVOC and health in non-industrial indoor environments. *Indoor Air* **1997**, *7*, 78–92. [CrossRef]

21. World Commission on Environment and Development (WCED). *Our Common Future*; Oxford University Press: Oxford, UK, 1987.

22. Huang, K.C.; Chiang, C.M.; Lee, C.C.; Cheng, Y.L.; Lin, W.T. Validation of the reduction indoor formaldehyde concentrations for sorptive test system by green building material in Taiwan. *J. Archit.* **2012**, *80*, 63–83. (In Chinese)

environments

MDPI

Review

Particulate Matter from the Road Surface Abrasion as a Problem of Non-Exhaust Emission Control

Magdalena Penkała [1], Paweł Ogrodnik [2] and Wioletta Rogula-Kozłowska [2,3,*]

[1] The State School of Higher Education, 54 Pocztowa St., 22-100 Chełm, Poland; mmagdapenkala@gmail.com
[2] The Main School of Fire Service, Faculty of Fire Safety Engineering, 52/54 Słowackiego St., 01-629 Warsaw, Poland; pogrodnik@sgsp.edu.pl
[3] Institute of Environmental Engineering, Polish Academy of Sciences, 34 M. Skłodowskiej-Curie St., 41-819 Zabrze, Poland
*   Correspondence: wrogula@sgsp.edu.pl or wioletta.rogula-kozlowska@ipis.zabrze.pl; Tel.: +48-225-617-553

Received: 27 October 2017; Accepted: 5 January 2018; Published: 7 January 2018

Abstract: Along with house heating and industry, emissions from road traffic (exhaust and tire, brake, car body or road surface abrasions) are one of the primary sources of particulate matter (PM) in the atmosphere in urban areas. Though numerous regulations and vehicle-control mechanisms have led to a significant decline of PM emissions from vehicle exhaust gases, other sources of PM remain related to road and car abrasion are responsible for non-exhaust emissions. Quantifying these emissions is a hard problem in both laboratory and field conditions. First, we must recognize the physicochemical properties of the PM that is emitted by various non-exhaust sources. In this paper, we underline the problem of information accessibility with regards to the properties and qualities of PM from non-exhaust sources. We also indicate why scarce information is available in order to find the possible solution to this ongoing issue.

Keywords: ambient particulate matter; street dust; exhaust vs. non-exhaust emission; heavy metals; health hazard

1. Introduction

Road traffic emissions caused by both exhaust and non-exhaust sources contribute significantly to the particulate matter (PM) concentration in an urban atmosphere [1–22]. Additionally, very fine particles that are emitted through various road traffic-related processes (e.g., brake wear), can penetrate the human organs [23–25]. A knowledge of physicochemical properties and PM sources is crucial for determining the environmental effects of PM [26–30].

Most of the PM mass, where PM is created by the abrasion of a car body, tires, or a road surface, is composed of particles with diameter ranging between 1 and 10 μm [9,31–35]. For this reason, coarse PM (fraction of atmospheric particles with an aerodynamic diameter in the 2.5–10 μm range; coarse PM) usually makes up most of the PM mass close to crossroads. In Switzerland, the PM_{10} in an urban street canyon may result from brake-disc abrasion (21%), resuspension (38%), or exhaust emissions (41%), while the PM_{10} found along a freeway consists mainly of resuspension dust (56%) and particles emitted by exhaust emissions (41%) [36]. Both the $PM_{2.5}$ (fraction of atmospheric particles with an aerodynamic diameter not exceeding 2.5 μm; fine PM) mass share of PM and the ambient concentration of $PM_{2.5}$ in the vicinities of roads are strongly affected by exhaust emissions than by non-exhaust emissions [11,12,20,37]. When compared to $PM_{2.5}$ from other areas, the effect of car fume emissions can mainly be seen in an increase in soot and some organic compounds in the $PM_{2.5}$ mass [3,4,6,11–13,20,21].

Most of the PM mass from non-exhaust emissions, regardless of the type of traffic site (road, crossroad, or highway), is made of resuspension dust. This is a mixture of particles derived from car body abrasions, brake pads, brake discs, tires, and road surface; and, on the other hand, from soil and particle elements that have settled on the surface of the road (originating from sources other than road traffic) [9,11,12,33,34]. These particles are usually enriched with various organic compounds [15,20–22,38]. In order to reduce the impact of resuspension dust on air quality in urban areas, appropriate methods are employed. For instance, for the prevention of PM resuspension, hardening, unpaved roads may be exposed to chemical-binder agents, as well as water-spraying in dry periods [39–41]. Frequently cleaning the surface via washing and sweeping is a practice used for paved roads. The reduction of PM emissions can also be achieved by limiting both vehicle speed and mass or by using a noise barrier on the roads (e.g., a green-wall sound barrier). The effectiveness of applied methods varies depending the frequency at which particles are removed and local meteorological conditions. The reduction of resuspension dust can reach an efficiency of up to 90% [39–41].

In this work, we demonstrate that the isolation of particles that only results from road surface abrasion from resuspension dust particles is particularly difficult. However, this step is necessary for the proper quantification of non-exhaust emissions in a particular area.

2. Non-Exhaust PM Emissions and Their Relation to Air Pollution in the Vicinity of Roads and Crossroads

Non-exhaust sources of PM include:

- Tires: PM from tire abrasions comprises among others, metals: Cd, Cu, Pb, Zn [33,42–44] and organic compounds such as natural rubber copolymer, organotin compounds, and soot [9].
- Brakes: PM from brake-pad and brake-disc abrasion consists of metals: Zn, Cu, Ti, Fe, Cu, Pb [33,45–51] and other specific compounds such as sulfate silicate, barium sulfate, carbon fibers, and graphite [9,52,53].
- The car body: some particles from the vehicle's consumable parts are released into the air; they may contain small amounts of metal like Zn and Fe [35].
- The road surface: PM from road-surface erosion containing characteristic compounds such as bitumen, cement, and resins [33].
- Paints: the composition of the paint intended for road-surface painting suggests that PM from this source may contain Pb and Ti [54].

It is clear that the level of PM air pollution resulting from non-exhaust sources at roads is dependent on traffic, speed as well as the shape and system of the road interchanges.

Released particles are permanently mixed coming from combustion, industry, exhaust emissions and soil [9,55–59]. PM from anthropogenic sources include particles from so-called municipal emissions (mainly soot, organic matter, and inorganic salts resulting from the combustion of coal and biomass in domestic furnaces, local coal-fired boiler plants, or heat plants [20,60–62]) and from industrial emissions (mainly fine particles enriched in heavy metals or persistent organic pollutants [63–68]). In addition, particles of salt, sand (coarse PM), or a mixture of both are released into the urban atmosphere, especially during the winter season [9,25,35,40,69]. In built-up areas, e.g., city centers, compact residential and service buildings close to traffic arteries significantly reduce air-mass exchange; this results in the accumulation of PM in the ground atmosphere layer [11,12,19,55,56]. The non-exhaust-based particles and particles from other sources creates complex chemical PM mixtures near traffic sites [11–13,21,33,35].

PM-bound elements, including some toxic metals, are subjected to similar phenomena. The ambient concentration of PM-bound elements in the areas affected by road-traffic emissions also depends on the traffic intensity, the vehicle-fleet characteristics car type and speed, the type of road surface, the road-cleaning intensity, and the concentration of PM components in the so-called urban background [70–73]. Given that, it is easy to imagine the difficulty in quantitatively dividing PM-bound elements, located in urban sites that are influenced by traffic emissions, into specific source groups.

Meanwhile, it is extremely important to clearly distribute these element mixtures as such a distribution constitutes the first and most important step in programs and scenarios that aim to reduce PM concentrations in areas characterized by high ambient concentrations of PM. The physical removal of elements from road and street surfaces via washing and sweeping does not sufficiently limit human exposure to PM resulting from traffic emission [55]. The complex process of overlapping the so-called urban background onto road traffic pollutants, along with the associated physical changes and chemical reactions, lead to the conclusion that PM close to roads may be much more toxic or may have a higher carcinogenic potential than PM outside such areas [11,12,20,70,74–77].

Using data derived from several European cities, Querol et al. [78] estimated that PM emissions from road traffic was roughly distributed between exhaust and non-exhaust emission sources. Similarly, in Berlin, Lenschow et al. [79] showed that in areas where air quality is determined by road traffic emissions (on urban roads) half of the PM_{10} mass comes from non-exhaust sources. Studies conducted in Poland in the last decade clearly showed that PM emissions from non-exhaust sources are significantly more present than PM from exhaust emissions at roads and crossroads [11–13,37]. This contribution was apparent even during smog episodes, when emissions from the combustion of fossil fuels in domestic furnaces determined the air quality in the city of Zabrze [19].

In this respect, the contribution of PM emissions from non-exhaust sources is large, certainly much larger than was previously thought. Nevertheless, we still lack detailed information [75,80–83]. Data and rationales are missing for the reliable subdivision of PM emissions into those from exhaust vs. those from non-exhaust sources, for the separation of road-traffic related PM from PM related to other sources [20,62], and, above all, for the segmentation of PM generally derived from non-exhaust sources into specific source groups (tire abrasion, brakes, road surface, etc.).

The simplest way of subdividing road-traffic related PM (though this does not apply to PM-bound elements) into specific sources is to assign a particular PM size, or rather particles belonging to a corresponding size range (i.e., size fractions), to specific sources; these compartments are characterized by so-called lower and upper limit cut-off diameters [84–86]. Many research papers are devoted to investigating the fractional composition of particles related to specific road-traffic sources. In general we can say that non-exhaust emissions mainly include coarse PM, while PM from exhaust emissions consists of fine particles belonging to $PM_{2.5}$ [34,87–89]. If we consider size when examining the distribution of PM particles in non-exhaust emission sources, particles of worn road surfaces usually belong to the coarse PM, while particles from the tires and brake discs are both coarse and fine [90]. In northern European countries, where various methods are used during winter to prevent icing on roads, and winter tires and snow chains are commonly used, $PM_{2.5-10}$ accounts for up to 90% of PM in atmospheric air [73,91]. However, brake discs were selected to be examined, as their usage is the most detailed, in terms of the amounts of emitted particles (emission factors) and their properties [9,33,55,80,87,92,93]. The identification and quantification of PM from other non-exhaust sources are much more difficult. The list of published papers describing the PM emission factors and properties of PM emitted by these sources is therefore much shorter. For example, it was shown that particles with a diameter ranging from several hundred nanometers to several tens of micrometers are produced during the brake-lining and disc-abrasion processes [33]. On the other hand, there are scientific reports showing that non-exhaust emissions (including brake wear) also produce nanoparticles [23–25]. The size of emitted particles depends primarily on the physical properties, shape and structure, and the chemical composition of the abrasive/erosion material, on the nature, value, and complexity of the forces acting on the material, and on the temporal and spatial variability of these parameters [94]. It should be clearly stated that the above-mentioned assignment of specific PM emission sources to appropriate size ranges is contractual and far from sufficient for separating PM into specific emission sources.

Rather than using particle-size distribution, a far more reliable way of achieving this separation involves examining the chemical composition (relatively elemental composition) of PM particles and comparing it with the chemical compositions of particles directly emitted by each of the individual emission sources [52,95–97]. These were widely described as so-called elemental PM profiles from both exhaust and non-exhaust emissions. Prior to the introduction of a law prohibiting the use of leaded petrol, Pb was a PM marker related to the combustion of gasoline in engines [98]. PM markers from working catalytic converters are metals from the platinum group such as Rh and Pt [99–101]. Zn, Cu, and Ti in urban air are often related to brake-pad erosion [33,47]; particles from brake-disc wear are typically characterized by Fe, Cu, Pb, and Zn [33,45,46,48–51]. To identify PM from tire abrasion, Cd, Cu, Pb, and Zn may be monitored, among others [33,42]. Occasionally, metals such as Fe and Zn are also linked to the corrosion of car-body parts [35].

The most difficult data to find involves the elemental composition of PM arising from the abrasion of road surfaces. The root of the problem lies in the nomenclature given to the material emitted directly by the worn road surface and to the material laid on road and then resuspended. Often, the same particles are attributed to several sources [20,33,62,102].

Fauser (1999) identified asphaltenes and maltenes as potential tracers for road-surface wear [103]. Kupiainen et al. (2003) used the presence of hornblende as a tracer for PM derived from road-surface wear [44]. However, this was in a controlled laboratory test, where the road-surface material constituted the only possible source of this mineral. Polycyclic aromatic hydrocarbons (PAHs) have been identified in bitumen samples, although the concentrations of individual PAHs are very low due to their removal during the distillation process [104]. The lack of a PAH compound unique to bitumen makes such compounds unusable as marker species for road-surface wear [105]. A number of metals have been detected in road bitumen samples including V, Ni, Fe, Mg, and Ca [48,106]. However, a comparison of concentrations in road bitumen with raw bitumen samples revealed much higher concentrations of these metals in the former, indicating that the road-surface material had incorporated these elements from other sources [48]. In conclusion, it may prove very difficult to identify suitable tracer species for road-surface wear [33].

It seems that we lack reliable information on the PM or dust emitted by different road surfaces' erosion processes, and this constitutes a significant gap in the current state of knowledge of non-exhaust sources of road traffic emissions. Such information should include: (i) the fractional composition data (ii) the chemical or elemental composition of the PM /dust particles emitted during the erosion of different road surfaces.

3. Types of Road Surfaces and Their Importance in the Non-Exhaust Emissions of PM

Road pavements can be classified by their construction, deformability, load-bearing capacity, and material types used for the driving layer. Due to the surfaces' deformability, they can be divided as follows:

- susceptible: surfaces with a structure that deforms plastically under the influence of loads (sett, gravel, and bitumen surfaces located on susceptible substrates).
- semi-rigid: asphalt surfaces with a foundation made of concrete, lean concrete, aggregates, or stabilized soils.
- rigid: surfaces with a structure that deforms elastically under a load (cement concrete surfaces) [107–109].

The surface's wearing course can be made of the following mineral-asphalt mixtures:

- asphalt concrete (AC), stone mastic asphalt (SMA), mastic asphalt (MA), very thin-layered asphalt concrete (BBTM), and porous asphalt (PA) [110].

A topcoat made of cement concrete can be made in the form of roofed, doweled and anchored, or reinforced plates [111].

Even without any measurements and tests, it is generally recognized that vehicles moving on unpaved roads will generate a significantly greater amount of PM than paved or hardened roads. In particular, the generation of high levels of dust is linked to ground and gravel made of crushed debris and slag roads. In comparison to paved roads (stone-paved, concrete, or asphalt roads), the top layers of unpaved roads (especially particles with a small diameter) move, as a result of the friction of the wheels on the surface. Loose surface grains move under the mechanical stress of the wheels. The free material is lifted under the effect of the vacuum that is generated by the tire surface breaking off from the road surface. Equally, dust is raised from the road surface onto the wheel tread [35,112,113]. However, concerning the impact on air quality unpaved roads are less interesting than paved roads. First of all, the chemical composition of the PM or dust particles emitted by unpaved roads is well known, and it is equal to the composition of soil in the case of forests for example, or to gravel and sand in the case of gravel roads for example [114]. Consequently, in terms of chemical properties, the PM or dust emitted during the erosion of such roads is almost identical, in a given area, with soil particles or the top soil layer [33]. The content in such particles may depend only on the degree of contamination of the soil or sand. Therefore, it basically depends on the road localization [115]. Road dust and roadside soil often contain metals, including Pb, Cu, Cd, and Zn, indicative of contamination by road traffic emissions. Through the calculation of crustal enrichment factors (CEFs), the presence of contaminants in road dust and roadside soil provides a means by which particles arising from anthropogenic sources can be separated from natural or crustal sources [11,12]. However, comparing the contribution of road-surface wear to that of resuspended road dust will require the development of an alternative approach [33]. Additionally, unpaved roads are usually located outside urban centers, in less-populated areas such as villages, small municipalities, and around small housing estates (access roads). They are therefore of little importance when compared to the state of air quality in city centers, where PM concentrations matter to authorities and sanitary services. Furthermore, it seems that the impact of unpaved-road emissions on the quality of atmospheric air is difficult and, almost impossible to estimate due to the high variability in time [116].

This is not the case for paved roads. First of all, they are built from mixes that are produced through strictly-defined processes. Their composition must therefore differ from the composition of natural components such as soil or sand. In addition, they are made of road surfaces where vehicles are able to move at high speeds, and varied with types ranging from compact passenger cars to several-ton trucks on highways. It seems, therefore, that the size of PM or dust particles generated during erosion may fit within a wide range Their composition determined by the structure of building material's original mixture may also vary depending on the place (road age, renovations, degree of wear, etc.) [116,117].

Most paved road surfaces have the same base: a mixture of aggregates such as bitumen or cement with different grain sizes, and modifiers such as fillers and binders [35]. The choice and proportions of the ingredients dictate the exact differences in chemical composition of mixtures made for the production of road surfaces [118]. Road surfaces can be widely classified as being made of either concrete or asphalt. Asphalt consists mainly (~95% of the mass) of mineral aggregates and various geological materials [93]. Its remaining content is mostly bituminous binder, modified by the addition of fillers and adhesives [106]. The bitumen contains many thousands of high-molecular-weight organic compounds (about 500–50,000 u), most of which are aliphatic and aromatic hydrocarbons [33].

As a result of increasing requirements of road surfaces' strength parameters, the continuous growth in their heavy traffic load, and the fact that extreme weather conditions occur at various latitudes, diverse types of modifiers are introduced into the bitumen-bonding compound or asphalt mix. This aims to improve both the properties of the road surface and the bond between the binder and mineral components. Adhesives are joined to aggregates using fillers and reinforcing fibers such as glass, fly ash, and shredded used tires [33,119]. This solution is increasingly prevalent due to problems related to the utilization of tires. Polymers, epoxy resins, and low-carbon steel are used as modifiers, while sulfur is used to increase the rigidity of the binder [33].

Concrete pavements are made of mineral aggregates, sand, and cement [35]. The literature provides little information on the chemical composition of concretes that are used to produce roads and on the dust emitted by such surfaces. This is related to the fact that there are many ways of choosing the composition and proportion of materials used to make a concrete mix. As a consequence there is no universal molecular formula for the concrete mix and therefore for particles emitted during the erosion of such road surfaces [33].

When concrete pavements are made, many different additives affect the strength of the material when they are introduced into the mix. Due to the fact that the structures are designed for a maximum period of 30 years, old concrete surfaces are significantly damaged and therefore not always adapted to heavy traffic and vehicle loads. This may result in emissions of PM and dust into the atmosphere, not only from the base-surface components released during the vehicle's movement but also from the processes and materials used for the surfaces' temporary repairs [120].

In order to promote the use of cement and concrete in the transport infrastructure, the non-profit European Concrete Paving Association (EUPAVE) was established in 2007. Their mission is to advocate and enable the wider use of cement and concrete applications in European transport infrastructures. They are involved in engagement with EU, national, and local decision makers, disseminating technical know-how, and presenting the benefits of using this material. According to EUPAVE, the use of cement and concrete in road-building implements a circular economy. Considering the hierarchy of waste management, in accordance with the European Commission's recommendations (prevention, reuse, recycling, recovery, and utilization), it seems obvious that construction using concrete, adheres to these principles. Concrete pavements have always been valued for their durability and low-maintenance costs, which are simple ways of obtaining ecological benefits [120].

Surface type also determines the wear of vehicles that move on it [35] The volume of dust emissions from used car tires in Britain in 1996 was 5.3×10^7 kg, while in Japan it was 2.1×10^8 kg in 2001. Within a year, Germany's emissions range between 55 and 657 kg·km^{-1}, depending on the type of road [121]. The asphalt surface causes less tire abrasion than concrete surfaces. In Arizona (in the United States), PM emissions are 1.4 to 2 times lower for asphalt roads than for concrete roads [9,122]. However, it is important to note what type of tires are used. Numerous studies have shown that winter tires generate much more dust, and this is mainly due to greater friction on the ground compared to their summer equivalents [9,35,123–126]. Furthermore, the use of metal constructions on tires, in the form of pins or chains, influences an increase in the level of road-surface abrasion [9,44,127]. In Sweden, studded tires are approved for use from October to April. The emission factor for PM$_{10}$ from tire abrasion close to roads and junctions is clearly higher during this period than in other months [127].

Finally, the emission of particulate matter by road-surface abrasion is difficult to isolate in PM and dust tests in the field, and such test, which aim to define its characteristics, are therefore performed under controlled laboratory conditions [36,44,125,128,129]. In many countries, practically no road with a wearing course of cement-based concrete existed until the second half of the 20th century. When this technology was reintroduced in the mid-1990s, there were many new modifications in the production of concrete. For example, some special additives, when introduced into the mix (e.g., silica), affect the surface strength. Because the surfaces are designed to last for about 30 years, there is now a noticeable increase in damaged old concrete surfaces, which are not always adapted to today's increase in traffic and vehicle loads. This can cause significant emissions of PM or dust into the atmosphere from both base-surface elements released during vehicle movement and materials used for the emergency repair of these surfaces [119].

4. Conclusions

Current research shows that direct road-surface abrasion is of minor importance when the road is undamaged and that emissions of PM from direct road-surface abrasion are then significantly lower than emissions from other sources of road dust like abrasion of car body, brakes or tires [36,44,125,128–130]. Nevertheless, data on this subject is clearly lacking. This scarcity concerns both the availability of data on PM/dust emissions from road-surface abrasion in various places, and studies on PM/dust emission factors and related elements from different types of road surfaces as current research were restricted to asphalt surface only. The return to the technology of road construction and renovation with concrete in recent years is more challenging in terms of PM mass, composition and size ranges emitted from abrasion roads in non-exhaust emission sources than results from current studies. The amounts of PM (emission factors) created during abrasion of concrete surfaces for different types of vehicles are not known, what is more, a chemical composition of particles released during this process is also unknown. The constant concrete upgrades aimed at their mechanical improvement can lead to the presence of dangerous elements in the environment. The increasing production of silica, used in concrete mixtures, could be a good example. Thus, more extensive research to establish emission factors and chemical composition of PM emitted from currently used road pavements are necessary. Such studies should be conducted under laboratory, controlled conditions for defining emission factors and chemical composition of PM (depending on vehicle type, speed, environmental conditions). On the other hand, analyzing the impact of abrasion emission reflecting realistic conditions is also very important. Defining how such emissions quantitatively and qualitatively change the character of PM near roads is crucial as it allows researchers also to assess PM parameters determining the power and scale of its influence in the environment. Gathering such information will allow for the revision of forecasts and scripts of PM impact on the environment and human health.

Acknowledgments: The work was financed in the frame of the Statutory Studies conducted in the Faculty of Fire Safety Engineering, The Main School of Fire Service, Poland.

Author Contributions: All authors conceived and wrote the paper.

Conflicts of Interest: The authors declare no conflict of interest.

References

1. Buckeridge, D.L.; Glazier, R.; Harvey, B.J.; Escobar, M.; Amrhein, C.; Frank, J. Effect of motor vehicle emissions on respiratory health in an urban area. *Environ. Health Perspect.* **2002**, *110*, 293–300. [CrossRef] [PubMed]
2. Fan, Z.; Meng, Q.; Weisel, C.; Shalat, S.; Laumbach, R.; Ohman-Strickland, P.; Black, K.; Rodriguez, M.; Bonanno, L. Acute Short-Term Exposures to $PM_{2.5}$ Generated by Vehicular Emissions and Cardiopulmonary Effects in Older Adults. *Epidemiology* **2006**, *17*, 213–214. [CrossRef]
3. Harrison, R.M.; Jones, A.M.; Lawrence, R.G. Major component composition of PM_{10} and $PM_{2.5}$ from roadside and urban background sites. *Atmos. Environ.* **2004**, *38*, 4531–4538. [CrossRef]
4. Harrison, R.M.; Smith, D.J.T.; Luhana, L. Source Apportionment of Atmospheric Polycyclic Aromatic Hydrocarbons Collected from an Urban Location in Birmingham, U.K. *Environ. Sci. Technol.* **1996**, *30*, 825–832. [CrossRef]
5. Health Effects Institute. *HEI Panel on the Health Effects of Traffic-Related Air Pollution: A Critical Review of the Literature on Emissions, Exposure, and Health Effects*; HEI Special Report 17; Health Effects Institute: Boston, MA, USA, 2010.
6. Hueglin, C.; Gehrig, R.; Baltensperger, U.; Gysek, M.; Monn, C.; Vonmont, H. Chemical characterization of $PM_{2.5}$, PM_{10} and coarse particles at urban, near-city and rural sites in Switzerland. *Atmos. Environ.* **2005**, *39*, 637–651. [CrossRef]
7. Karagulian, F.; Belis, C.A.; Dora, C.F.C.; Prüss-Ustün, A.M.; Bonjour, S.; Adair-Rohani, H.; Amann, M. Contributions to cities' ambient particulate matter (PM): A systematic review of local source contributions at global level. *Atmos. Environ.* **2015**, *120*, 475–483. [CrossRef]
8. Masiol, M.; Hofer, A.; Squizzato, S.; Piazza, R.; Rampazzo, G.; Pavoni, B. Carcinogenic and mutagenic risk associated to airborne particle-phase polycyclic aromatic hydrocarbons: A source apportionment. *Atmos. Environ.* **2012**, *60*, 375–382. [CrossRef]
9. Pant, P.; Harrison, R.M. Estimation of the contribution of road traffic emissions to particulate matter concentrations from field measurements: A review. *Atmos. Environ.* **2013**, *77*, 78–97. [CrossRef]
10. Rissler, J.; Swietlicki, E.; Bengtsson, A.; Boman, C.; Pagels, J.; Sandström, T.; Blomberg, A.; Löndahl, J. Experimental determination of deposition of diesel exhaust particles in the human respiratory tract. *J. Aerosol Sci.* **2012**, *48*, 18–33. [CrossRef]
11. Rogula-Kozłowska, W. Chemical composition and mass closure of ambient particulate matter at a crossroads and a highway in Katowice, Poland. *Environ. Prot. Eng.* **2015**, *41*, 15–29.

12. Rogula-Kozłowska, W. Traffic-Generated Changes in the Chemical Characteristics of Size-Segregated Urban Aerosols. *Bull. Environ. Contam. Toxicol.* **2014**, *93*, 493–502. [CrossRef] [PubMed]
13. Rogula-Kozłowska, W.; Pastuszka, J.S.; Talik, E. Influence of Vehicular Traffic on Concentration and Particle Surface Composition of PM_{10} and $PM_{2.5}$ in Zabrze, Poland. *Pol. J. Environ. Stud.* **2008**, *17*, 539–548.
14. Enroth, J.; Saarikoski, S.; Niemi, J.; Kousa, A.; Ježek, I.; Močnik, G.; Carbone, S.; Kuuluvainen, H.; Rönkkö, T.; Hillamo, R.; et al. Chemical and physical characterization of traffic particles in four different highway environments in the Helsinki metropolitan area. *Atmos. Chem. Phys.* **2016**, *16*, 5497–5512. [CrossRef]
15. Geller, M.D.; Ntziachristos, L.; Mamakos, A.; Samaras, Z.; Schmitz, D.A.; Froines, J.R.; Sioutas, C. Physicochemical and redox characteristics of particulate matter (PM) emitted from gasoline and diesel passenger cars. *Atmos. Environ.* **2006**, *40*, 6988–7004. [CrossRef]
16. Harrison, R.M.; Tilling, R.; Callén Romero, M.S.; Harrad, S.; Jarvis, K. A study of trace metals and polycyclic aromatic hydrocarbons in the roadside environment. *Atmos. Environ.* **2003**, *37*, 2391–2402. [CrossRef]
17. Mauderly, J.L. Toxicological and epidemiological evidence for health risks from inhaled engine emissions. *Environ. Health Perspect.* **1994**, *102*, 165–171. [CrossRef] [PubMed]
18. Pant, P.; Shi, Z.; Pope, F.D.; Harrison, R.M. Characterization of Traffic-Related Particulate Matter Emissions in a Road Tunnel in Birmingham, UK: Trace Metals and Organic Molecular Markers. *Aerosol Air Qual. Res.* **2017**, *17*, 117–130. [CrossRef]
19. Pastuszka, J.S.; Rogula-Kozłowska, W.; Zajusz-Zubek, E. Characterization of PM_{10} and $PM_{2.5}$ and associated heavy metals at the crossroads and urban background site in Zabrze, Upper Silesia, Poland, during the smog episodes. *Environ. Monit. Assess.* **2010**, *168*, 613–627. [CrossRef] [PubMed]
20. Rogula-Kozłowska, W.; Kozielska, B.; Klejnowski, K. Concentration, Origin and Health Hazard from Fine Particle-Bound PAH at Three Characteristic Sites in Southern Poland. *Arch. Environ. Contam. Toxicol.* **2013**, *91*, 349–355. [CrossRef] [PubMed]
21. Rogula-Kopiec, P.; Kozielska, B.; Rogula-Kozłowska, W. Road Traffic Effects in Size-segregated Ambient Particle-bound PAHs. *Int. J. Environ. Res.* **2016**, *10*, 531–542.
22. Rogula-Kozłowska, W.; Rogula-Kopiec, P.; Klejnowski, K.; Blaszczyk, J. Influence of vehicular traffic on concentration and mass size distribution of two fractions of carbon in an urban area atmospheric aerosol. *Rocz. Ochr. Śr.* **2013**, *15*, 1623–1644.
23. Grigoratos, T.; Martini, G. Brake wear particle emissions: A review. *Environ. Sci. Pollut. Res.* **2015**, *22*, 2491–2504. [CrossRef] [PubMed]
24. Grigoratos, T.; Martini, G. *Non-Exhaust Traffic Related Emissions. Brake and Tyre Wear PM. JRC Science and Policy Reports. Literature Review*; European Commission Joint Research Centre Institute of Energy and Transport: Luxembourg, 2014.
25. Kumar, P.; Pirjola, L.; Ketzel, M.; Harrison, R.M. Nanoparticle emissions from 11 non-vehicle exhaust sources—A review. *Atmos. Environ.* **2013**, *67*, 252–277. [CrossRef]
26. Badyda, A.J.; Dąbrowiecki, P.; Czechowski, P.O.; Majewski, G.; Doboszyńska, A. Traffic-Related Air Pollution and Respiratory Tract Efficiency. *Adv. Exp. Med. Biol.* **2014**, *834*, 31–38.
27. Badyda, A.J.; Dąbrowiecki, P.; Czechowski, P.O.; Majewski, G. Risk of bronchi obstruction among non-smokers—Review of environmental factors affecting bronchoconstriction. *Respir. Physiol. Neurobiol.* **2014**, *209*, 39–46. [CrossRef] [PubMed]
28. De Kok, T.M.C.M.; Driece, H.A.L.; Hogervorst, J.G.F.; Briedé, J.J. Toxicological assessment of ambient and traffic-related particulate matter: A review of recent studies. *Mutat. Res.* **2006**, *613*, 103–122. [CrossRef] [PubMed]
29. Grynkiewicz Bylina, B.; Rakwic, B.; Pastuszka, J.S. Assessment of Exposure to Traffic-Related Aerosol and to Particle-Associated PAHs in Gliwice, Poland. *Pol. J. Environ. Stud.* **2005**, *14*, 117–123.
30. Schauer, J.J.; Lough, G.C.; Shafer, M.M.; Christensen, W.F.; Arndt, M.F.; Deminter, J.T.; Park, J.S. Characterization of metals emitted from motor vehicles. *Res. Rep. Health Eff. Inst.* **2006**, *133*, 77–88.
31. Geller, M.D.; Sardar, S.B.; Phuleria, H.; Fine, P.M.; Sioutas, C. Measurements of Particle Number and Mass Concentrations and Size Distributions in a Tunnel Environment. *Environ. Sci. Technol.* **2005**, *39*, 8653–8663. [CrossRef] [PubMed]
32. Kwak, J.-H.; Kim, H.; Lee, J.; Lee, S. Characterization of non-exhaust coarse and fine particles from on-road driving and laboratory measurements. *Sci. Total Environ.* **2013**, *458–460*, 273–282. [CrossRef] [PubMed]

33. Thorpe, A.J.; Harrison, R.M. Sources and properties of non-exhaust particulate matter from road traffic: A review. *Sci. Total Environ.* **2008**, *400*, 270–282. [CrossRef] [PubMed]

34. Thorpe, A.J.; Harrison, R.M.; Boulter, P.G.; Mccrae, I.S. Estimation of particle resuspension source strength on a major London Road. *Atmos. Environ.* **2007**, *41*, 8007–8020. [CrossRef]

35. Warner, L.R.; Sokhi, R.; Luhana, L.; Boulter, P.G. Non-Exhaust Particle Emissions from Road Transport: A Literature Review. Unpublished Report PR/SE/213/00. 2001. Available online: http://lat.eng.auth.gr/particulates/old_website/eterg/files/PR_SE_~1.PDF (accessed on 27 October 2017).

36. Bukowiecki, N.; Lienemann, P.; Hill, M.; Furger, M.; Richard, A.; Amato, F.; Prévôt, A.S.H.; Baltensperger, U.; Buchmann, B.; Gehrig, R. PM$_{10}$ emission factors for non-exhaust particles generated by road traffic in an urban street canyon and along a freeway in Switzerland. *Atmos. Environ.* **2010**, *44*, 2330–2340. [CrossRef]

37. Rogula-Kozłowska, W.; Pastuszka, J.S.; Talik, E. *Właściwości aerozolu ze źródeł komunikacyjnych*; IPIŚ PAN: Zabrze, Poland, 2011; Volume 80, p. 111.

38. Oyama, B.S.; Andrade, M.; Herckes, P.; Dusek, U.; Röckmann, T.; Holzinger, R. Chemical characterization of organic particulate matter from on-road traffic in São Paulo, Brazil. *Atmos. Chem. Phys.* **2016**, *16*, 14397–14408. [CrossRef]

39. Dziugieł, M.; Bogacki, M. Metodyka wyznaczania emisji niezorganizowanej pyłu do powietrza z dróg oraz eksploatacji składowisk w kopalni odkrywkowej surowców mineralnych. *Przegląd Górniczy* **2013**, *12*, 68–74. (In Polish)

40. The Scientific Basis of Street Cleaning Activities as Road Dust Mitigation Measure, Action B7. 2013. Available online: http://airuse.eu/wp-content/uploads/2013/11/B7-3-ES_road-cleaning.pdf (accessed on 27 October 2017).

41. *Ravensworth Underground Mine-Coal Mine Particulate Matter Control Best Management Practice Determination*; Xstrata Coal: Zug, Switzerland, 1979.

42. Adamiec, E.; Jarosz-Krzemińska, E.; Wieszała, R. Heavy metals from non-exhaust vehicle emissions in urban and motorway road dusts. *Environ. Monit. Assess.* **2016**, *188*, 369. [CrossRef] [PubMed]

43. Pirjola, L.; Johansson, C.; Kupiainen, K.; Stojiljkovic, A.; Karlsson, H.; Hussein, T. Road Dust Emissions from Paved Roads Measured Using Different Mobile Systems. *J. Air Waste Manag.* **2010**, *60*, 1422–1433. [CrossRef]

44. Kupiainen, K.; Tervahattu, H.; Räisänen, M. Experimental studies about the impact of traction sand on urban road dust composition. *Sci. Total Environ.* **2003**, *308*, 175–184. [CrossRef]

45. Garg, B.D.; Cadle, S.H.; Mulawa, P.A.; Groblicki, P.J.; Laroo, C.; Parr, G.A. Brake Wear Particulate Matter Emissions. *Environ. Sci. Technol.* **2000**, *34*, 4463–4469. [CrossRef]

46. Hildemann, L.M.; Markowski, G.R.; Cass, G.R. Chemical composition of emissions from urban sources of fine organic aerosol. *Environ. Sci. Technol.* **1991**, *25*, 744–759. [CrossRef]

47. Iijima, A.; Sato, K.; Yano, K.; Tago, H.; Kato, M.; Kimura, H.; Furuta, N. Particle size and composition distribution analysis of automotive brake abrasion dusts for the evaluation of antimony sources of airborne particulate matter. *Atmos. Environ.* **2007**, *41*, 4908–4919. [CrossRef]

48. Kennedy, P.; Gadd, J. *Preliminary Examination of Trace Elements in Tyres, Brake Pads, and Road Bitumen in New Zealand*; Infrastructure Auckland: Auckland, New Zealand, 2003.

49. Kennedy, P.; Gadd, J.; Moncrieff, I. *Emission Factors for Contaminants Released by Motor Vehicles in New Zealand*; Infrastructure Auckland: Auckland, New Zealand, 2002.

50. Legret, M.; Pagotto, C. Evaluation of pollutant loadings in the runoff waters from a major rural highway. *Sci. Total Environ.* **1999**, *235*, 143–150. [CrossRef]

51. Westerlund, K.G.; Johansson, C. Emission of Metals and Particulate Matter Due to Wear of Brake Linings in Stockholm. *Air Pollut.* **2002**, *10*, 793–802.

52. Dongarrà, G.; Manno, E.; Varrica, D. Possible markers of traffic-related emissions. *Environ. Monit. Assess.* **2009**, *154*, 117–125. [CrossRef] [PubMed]

53. Ingo, G.M.; D'Uffizi, M.; Falso, G.; Bultrini, G.; Padeletti, G. Thermal and microchemical investigation of automotive brake pad wear residues. *Thermochim. Acta* **2004**, *418*, 61–68. [CrossRef]

54. Przybek, P. Materiały Malarskie—Pomoc Dydaktyczna. 2004. Available online: http://bianda.cba.pl/tppch/sload/materiały_malarskie.pdf (accessed on 9 August 2017).

55. Harrison, R.M.; Hester, R.E. *Environmental Impacts of Road Vehicles: Past, Present and Future*; Royal Society of Chemistry: London, UK, 2017.

56. Harrison, R.M.; Jones, A.M.; Barrowcliffe, R. Field study of the influence of meteorological factors and traffic volumes upon suspended particle mass at urban roadside sites of differing geometries. *Atmos. Environ.* **2004**, *38*, 6361–6369. [CrossRef]

57. Amato, F.; Schaap, M.; Denier van der Gon, H.A.C.; Pandolfi, M.; Alastuey, A.; Keuken, M.; Querol, X. Effect of rain events on the mobility of road dust load in two Dutch and Spanish roads. *Atmos. Environ.* **2012**, *62*, 352–358. [CrossRef]

58. Olszowski, T. Changes in PM_{10} concentration due to large-scale rainfall. *Arab. J. Geosci.* **2016**, *9*, 160. [CrossRef]

59. Amato, F.; Karanasiou, A.; Cordoba, P.; Alastuey, A.; Moreno, T.; Lucarelli, F.; Nava, S.; Calzolai, G.; Querol, X. Effects of Road Dust Suppressants on PM Levels in a Mediterranean Urban Area. *Environ. Sci. Technol.* **2014**, *48*, 8069–8077. [CrossRef] [PubMed]

60. Rogula-Kozłowska, W.; Błaszczak, B.; Szopa, S.; Klejnowski, K.; Sówka, I.; Zwoździak, A.; Jabłońska, M.; Mathews, B. $PM_{2.5}$ in the central part of Upper Silesia, Poland: Concentrations, elemental composition, and mobility of components. *Environ. Monit. Assess.* **2013**, *185*, 581–601. [CrossRef] [PubMed]

61. Rogula-Kozłowska, W.; Klejnowski, K.; Rogula-Kopiec, P.; Ośródka, L.; Krajny, E.; Błaszczak, B.; Mathews, B. Spatial and seasonal variability of the mass concentration and chemical composition of $PM_{2.5}$ in Poland. *Air Qual. Atmos. Health* **2014**, *7*, 41–58. [CrossRef] [PubMed]

62. Rogula-Kozłowska, W.; Majewski, G.; Czechowski, P.O. The size distribution and origin of elements bound to ambient particles: A case study of a Polish urban area. *Environ. Monit. Assess.* **2015**, *187*, 240. [CrossRef] [PubMed]

63. Klejnowski, K.; Rogula-Kozłowska, W.; Łusiak, T. Some metals and polycyclic aromatic hydrocarbons in fugitive PM_{10} emissions from the coking process. *Environ. Prot. Eng.* **2012**, *38*, 59–71.

64. Konieczyński, J.; Zajusz-Zubek, E. Distribution of selected trace elements in dust containment and flue gas desulphurisation products from coal-fired power plants. *Arch. Environ. Prot.* **2011**, *37*, 3–14.

65. Kozielska, B.; Konieczyński, J. Polycyclic aromatic hydrocarbons in dust emitted from stoker-fired boilers. *Environ. Technol.* **2007**, *28*, 895–903. [CrossRef] [PubMed]

66. Kozielska, B.; Konieczyński, J. Polycyclic aromatic hydrocarbons in particulate matter emitted from coke oven battery. *Fuel* **2015**, *144*, 327–334. [CrossRef]

67. Rogula-Kopiec, P.; Rogula-Kozłowska, W.; Kozielska, B.; Sówka, I. PAH Concentrations Inside a Wood Processing Plant and the Indoor Effects of Outdoor Industrial Emissions. *Pol. J. Environ. Stud.* **2015**, *24*. [CrossRef]

68. Zajusz-Zubek, E.; Konieczyński, J. Dynamics of trace elements release in a coal pyrolysis process. *Fuel* **2003**, *82*, 1281–1290. [CrossRef]

69. Dahl, A.; Gharibi, A.; Swietlicki, E.; Gudmundsson, A.; Bohgard, M.; Ljungman, A.; Blomgvist, G.; Gustafsson, M. Traffic-generated emissions of ultrafine particles from pavement-tire interface. *Atmos. Environ.* **2006**, *40*, 1314–1323. [CrossRef]

70. Amato, F.; Pandolfi, M.; Moreno, T.; Furger, M.; Pey, J.; Alastuey, A.; Bukowiecki, N.; Prevot, A.S.H.; Baltensperger, U.; Querol, X. Sources and variability of inhalable road dust particles in three European cities. *Atmos. Environ.* **2011**, *45*, 6777–6787. [CrossRef]

71. Amato, F.; Viana, M.; Richard, A.; Furger, M.; Prévôt, A.S.H.; Nava, S.; Lucarelli, F.; Bukowiecki, N.; Alastuey, A.; Reche, C.; et al. Size and time-resolved roadside enrichment of atmospheric particulate pollutants. *Atmos. Chem. Phys.* **2011**, *11*, 2917–2931. [CrossRef]

72. Duong, T.T.T.; Lee, B.K. Determining contamination level of heavy metals in road dust from busy traffic areas with different characteristics. *J. Environ. Manag.* **2011**, *92*, 554–562. [CrossRef] [PubMed]

73. Omstedt, G.; Bringfelt, B.; Johansson, C. A model for vehicle-induced non-tailpipe emissions of particles along Swedish roads. *Atmos. Environ.* **2005**, *39*, 6088–6097. [CrossRef]

74. Adachi, K.; Tainosho, Y. Characterization of heavy metal particles embedded in tire dust. *Environ. Int.* **2004**, *30*, 1009–1017. [CrossRef] [PubMed]

75. Denier van der Gon, H.A.C.; Gerlofs-Nijland, M.E.; Gehrig, R.; Gustafsson, M.; Janssen, N.; Harrison, R.M.; Hulskotte, J.; Johansson, C.; Jozwicka, M.; Keuken, M.; et al. The Policy Relevance of Wear Emissions from Road Transport, Now and in the Future—An International Workshop Report and Consensus Statement. *J. Air Waste Manag.* **2013**, *63*, 136–149. [CrossRef]

76. Hjortenkrans, D.S.T.; Bergbäck, B.G.; Häggerud, A.V. Metal Emissions from Brake Linings and Tires: Case Studies of Stockholm, Sweden 1995/1998 and 2005. *Environ. Sci. Technol.* **2007**, *41*, 5224–5230. [CrossRef] [PubMed]

77. Johansson, C.; Norman, M.; Burman, L. Road traffic emission factors for heavy metals. *Atmos. Environ.* **2009**, *43*, 4681–4688. [CrossRef]

78. Querol, X.; Alastuey, A.; Ruiz, C.R.; Artinano, B.; Hansson, H.C.; Harrison, R.M.; Buringh, E.; ten Brink, H.M.; Lutz, M.; Bruckmann, P.; et al. Speciation and origin of PM_{10} and $PM_{2.5}$ in selected European cities. *Atmos. Environ.* **2004**, *38*, 6547–6555. [CrossRef]

79. Lenschow, P.; Abraham, H.J.; Kutzner, K.; Lutz, M.; Preuß, J.D.; Reichenbacher, W. Some ideas about the sources of PM_{10}. *Atmos. Environ.* **2001**, *35*, 23–33. [CrossRef]

80. Amato, F.; Karanasiou, A.; Moreno, T.; Alastuey, A.; Orza, J.A.G.; Lumbreras, J.; Borge, R.; Boldo, E.; Linares, C.; Querol, X. Emission factors from road dust resuspension in a Mediterranean freeway. *Atmos. Environ.* **2012**, *61*, 580–587. [CrossRef]

81. Harrison, R.M.; Beddows, D.C.S.; Dall'Osto, M. PMF Analysis of Wide-Range Particle Size Spectra Collected on a Major Highway. *Environ. Sci. Technol.* **2011**, *45*, 5522–5528. [CrossRef] [PubMed]

82. Harrison, R.M.; Jones, A.M.; Gietl, J.; Yin, J.; Green, D.C. Estimation of the Contributions of Brake Dust, Tire Wear, and Resuspension to Nonexhaust Traffic Particles Derived from Atmospheric Measurements. *Environ. Sci. Technol.* **2012**, *46*, 6523–6529. [CrossRef] [PubMed]

83. Kuhlbusch, T.A.J.; John, A.C.; Quass, U. Sources and source contributions to fine particles. *Biomarkers* **2009**, *14*, 23–28. [CrossRef] [PubMed]

84. Friedlander, S.K. The characterization of aerosols distributed with respect to size and chemical composition—I. *J. Aerosol Sci.* **1970**, *1*, 295–307. [CrossRef]

85. Friedlander, S.K. The characterization of aerosols distributed with respect to size and chemical composition—II. *J. Aerosol Sci.* **1971**, *2*, 331–340. [CrossRef]

86. Hinds, W.C. *Aerosol Technology: Properties, Behavior, and Measurement of Airborne Particles*, 2nd ed.; John Wiley and Sons: New York, NY, USA, 1998.

87. Abu-Allaban, M.; Gillies, J.A.; Gertler, A.W.; Clayton, R.; Profitt, D. Tailpipe, resuspended road dust, and brake-wear emission factors from on-road vehicles. *Atmos. Environ.* **2003**, *37*, 5283–5293. [CrossRef]

88. Kam, W.; Liacos, J.W.; Schauer, J.J.; Delfino, R.J.; Sioutas, C. Size-segregated composition of particulate matter (PM) in major roadways and surface streets. *Atmos. Environ.* **2012**, *55*, 90–97. [CrossRef]

89. Tervahattu, H.; Kupiainen, K.J.; Räisänen, M.; Makela, T.; Hillamo, R. Generation of urban road dust from anti-skid and asphalt concrete aggregates. *J. Hazard. Mater.* **2006**, *132*, 39–46. [CrossRef] [PubMed]

90. Keuken, M.; Denier van der Gon, H.; van der Valk, K. Non-exhaust emissions of PM and the efficiency of emission reduction by road sweeping and washing in the Netherlands. *Sci. Total Environ.* **2010**, *408*, 4591–4599. [CrossRef] [PubMed]

91. Forsberg, B.; Hansson, H.C.; Johansson, C.; Areskoug, H.; Persson, K.; Järvholm, B. Comparative Health Impact Assessment of Local and Regional Particulate Air Pollutants in Scandinavia. *Ambio* **2005**, *34*, 11–19. [CrossRef] [PubMed]

92. Von Uexküll, O.; Skerfving, S.; Doyle, R.; Braungart, M. Antimony in brake pads-a carcinogenic component? *J. Clean. Prod.* **2005**, *13*, 19–31. [CrossRef]

93. Woodside, A.R. *Aggregates and Fillers in Asphalt Surfacings: A Guide to Asphalt Surfacings and Treatments Used for the Surface Course of Road Pavements*; The National Academies of Sciences, Engineering, and Medicine: Washington, DC, USA, 1998.

94. Watson, J.G.; Chow, J.C. Receptormodels for source apportionment of suspended particles. In *Introduction to Environmental Forensics*, 2nd ed.; Academic Press: New York, NY, USA, 2007; Volume 2, pp. 279–316.

95. Chow, J.C. Measurement Methods to Determine Compliance with Ambient Air Quality Standards for Suspended Particles. *J. Air Waste Manag.* **1995**, *45*, 320–382. [CrossRef]

96. Lawrence, S.; Sokhi, R.; Ravindra, K.; Mao, H.; Prain, H.D.; Bull, I.D. Source apportionment of traffic emissions of particulate matter using tunnel measurements. *Atmos. Environ.* **2013**, *77*, 548–557. [CrossRef]

97. Rybak, J. Accumulation of Major and Trace Elements in Spider Webs. *Water Air Soil Pollut.* **2015**, *226*, 105. [CrossRef] [PubMed]

98. Richter, P.; Griño, P.; Ahumada, I.; Giordano, A. Total element concentration and chemical fractionation in airborne particulate matter from Santiago, Chile. *Atmos. Environ.* **2007**, *41*, 6729–6738. [CrossRef]

99. Gómez, B.; Gómez, M.; Sanchez, J.L.; Fernandez, R.; Palacios, M.A. Platinum and rhodium distribution in airborne particulate matter and road dust. *Sci. Total Environ.* **2001**, *269*, 131–144. [CrossRef]
100. Heck, R.M.; Farrauto, R.J. Automobile exhaust catalysts. *Appl. Catal. A* **2001**, *221*, 443–457. [CrossRef]
101. Virtanen, A.; Keskinen, J.; Ristimäki, J.; Rönkkö, T.; Vaaraslahti, K. *Reducing Particulate Emissions in Traffic and Transport*; Views and Conclusions from the FINE Particles—Technology. Environment and Health Technology Programme; Tekes: Tampere, Finland, 2006.
102. AQEG. *Particulate Matter in the United Kingdom*; DEFRA: London, UK, 2005.
103. Fauser, P.; Tjell, J.C.; Bjerg, P.L. *Particulate air Pollution, with Emphasis on Traffic Generated Aerosols*; Riso National Laboratory and Technikal University of Denmark: Roskilde, Denmark, 1999.
104. Gadd, J.; Kennedy, P. *Preliminary Examination of Organic Compounds present in Tyres, Brake Pads, and Road Bitumen in New Zealand*, rev. ed.; Prepared for Ministry of Transport; Infrastructure Auckland: Auckland, New Zealand, 2003.
105. Boulter, P. *A review of Emission Factors and Models for Road Vehicle Non-Exhaust Particulate Matter*; Report Prepared for DEFRA; TRL: Wokingham, UK, 2005.
106. Lindgren, A. Asphalt wear and pollution transport. *Sci. Total Environ.* **1996**, *189–190*, 281–286. [CrossRef]
107. Piłat, J.; Radziszewski, P. *Nawierzchnie Asfaltowe*; Wydawnictwa Komunikacji I Łączności: Warsaw, Poland, 2007.
108. Dodds, C.J.; Robson, J.D. The description of road surface roughness. *J. Sound Vib.* **1973**, *31*, 1751–1783. [CrossRef]
109. Robson, J.D. Road surface description and vehicle response. *Int. J. Veh. Des.* **1979**, *1*. [CrossRef]
110. Katalog Typowych Konstrukcji Nawierzchni Podatnych I Półsztywnych. GDDKiA: Warsaw, Poland, 2014. Available online: https://www.gddkia.gov.pl/userfiles/articles/p/prace-naukowo-badawcze-po-roku-2_3432/Weryfikacja%20KataloguTNPiP_Etap4_final_11%2003%202013.pdf (accessed on 11 March 2013).
111. Katalog Typowych Konstrukcji Nawierzchni Sztywnych. GDDKiA: Warsaw, Poland, 2014. Available online: https://www.gddkia.gov.pl/userfiles/articles/z/zarzadzenia-generalnego-dyrektor_13901/zarzadzenie%2030%20zalacznik.pdf (accessed on 16 June 2016).
112. Gillies, J.A.; Etyemezian, V.; Kuhns, H.; Nikolic, D.; Gillette, D.A. Effect of vehicle characteristics on unpaved road dust emissions. *Atmos. Environ.* **2005**, *39*, 2341–2347. [CrossRef]
113. Gillies, J.A.; Watson, J.G.; Rogers, C.F.; DuBois, D.; Chow, J.C.; Langston, R.; Sweet, J. Long-Term Efficiencies of Dust Suppressants to Reduce PM_{10} Emissions from Unpaved Roads. *J. Air Waste Manag.* **1999**, *49*, 31–36. [CrossRef] [PubMed]
114. Watson, J.G.; Rogers, C.F.; Chow, J.C.; DuBois, D.; Gillies, J.A.; Derby, J.; Moosmüller, H. *Effectiveness Demonstration of Fugitive Dust Control Methods for Public Unpaved Roads and Unpaved Shoulders on Paved Roads*; Final Report DRI; DRI: Reno, NV, USA, 1996.
115. Christoforidis, A.; Stamatis, N. Heavy metal contamination in street dust and roadside along the main national road in Kavala's region, Greece. *Geoderma* **2009**, *151*, 257–263. [CrossRef]
116. Forman, R.T.T.; Sperling, D.; Bissonette, J.A.; Clevenger, A.P.; Cutshall, C.D.; Dale, V.H.; Fahrig, L.; France, R.L.; Goldman, C.R.; Heanue, K.; et al. *Road Ecology. Science and Solutions*; Island Press: Washington, DC, USA, 2013.
117. Kozłowski, W.; Surowiecki, A. Kierunki rozwoju konstrukcji nawierzchni dróg wiejskich. *Problemy Inżynierii Rolniczej* **2011**, *1*, 173–183.
118. Nicholls, J. *Asphalt Surfacings: A Guide to Asphalt Surfacings and Treatments Used for the Surface Course of Road Pavements*; Transport Research Laboratory, E&FN SPON An Imprint of Routledge: London, UK, 1998.
119. NIOSH. *Hazard Review: Health Effects of Occupational Exposure to Asphalt*; Department of Health and Human Services [DHHS] and NIOSH: Cincinnati, OH, USA, 2000.
120. Szruba, M. Nawierzchnie betonowe. *Nowoczesne Budownictwo Inżynieryjne* **2016**, *10*, 56–58.
121. Yongming, H.; Peixuan, D.; Junji, C.; Posmentier, E.S. Multivariate analysis of heavy metal contamination in urban dusts of Xi'an, Central China. *Sci. Total Environ.* **2006**, *355*, 176–186. [CrossRef] [PubMed]
122. Allen, J.O.; Alexandrova, O.; Kaloush, K.E. *Tire Wear Emissions for Asphalt Rubber and Portland Cement Concrete Pavement Surfaces*; Report Submitted to Arizona Department of Transportation; Arizona State University: Tempe, AZ, USA, 2006.
123. Gustafsson, M.; Blomqvist, G.; Gudmundsson, A.; Dahl, A.; Swietlicki, E.; Bohgard, M.; Lindbom, J.; Ljungman, A. Properties and toxicological effects of particles from the interaction between tyres, road pavement and winter traction material. *Sci. Total Environ.* **2008**, *393*, 226–240. [CrossRef] [PubMed]

124. Hussein, T.; Johansson, C.; Karlsson, H.; Hansson, H.C. Factors affecting non-tailpipe aerosol particle emissions from paved roads: On-road measurements in Stockholm, Sweden. *Atmos. Environ.* **2008**, *42*, 688–702. [CrossRef]
125. Kupiainen, K.J.; Tervahattu, H.; Räisänen, M.; Mäkelä, T.; Aurela, M.; Hillamo, R. Size and Composition of Airborne Particles from Pavement Wear, Tires, and Traction Sanding. *Environ. Sci. Technol.* **2005**, *39*, 699–706. [CrossRef] [PubMed]
126. Schaap, M.; Manders, A.M.M.; Hendriks, E.C.J.; Cnossen, J.M.; Segers, A.J.S.; Denier van der Gon, H.A.C.; Jozwicka, M.; Sauter, F.; Velders, G.; Matthijsen, J.; et al. *Regional Modelling of PM10 over the Netherlands*; Technical Report 500099008; Netherlands Environmental Assessment Agency, (PBL): Bilthoven, The Netherlands, 2009.
127. Sjodin, A.; Ferm, M.; Bjork, A.; Rahmberg, M.; Gudmundsson, A.; Swietlickli, E.; Johansson, C.; Gustafsson, M.; Blomqvist, G. *Wear Particles from Road Traffic: A Field, Laboratory and Modelling Study*; IVL Report B1830; IVL: Göteborg, Sweden, 2010.
128. Ferm, M.; Sjöberg, K. Concentrations and emission factors for $PM_{2.5}$ and PM_{10} from road traffic in Sweden. *Atmos. Environ.* **2015**, *119*, 211–219. [CrossRef]
129. Gustafsson, M.; Blomqvist, G.; Gudmundsson, A.; Dahl, A.; Jonsson, P.; Swietlicki, E. Factors influencing PM_{10} emissions from road pavement wear. *Atmos. Environ.* **2009**, *43*, 4699–4702. [CrossRef]
130. Gehrig, R.; Zeyer, K.; Bukowiecki, N.; Lienemann, P.; Poulikakos, L.D.; Furger, M.; Buchmann, B. Mobile load simulators—A tool to distinguish between the emissions due to abrasion and resuspension of PM_{10} from road surfaces. *Atmos. Environ.* **2010**, *44*, 4937–4943. [CrossRef]

environments

MDPI

Article

Pervious Concrete as an Environmental Solution for Pavements: Focus on Key Properties

Marek Kováč * and **Alena Sičáková** [ID]

Faculty of Civil Engineering, Technical University of Košice, 042 00 Košice, Slovakia; alena.sicakova@tuke.sk
* Correspondence: marek.kovac@tuke.sk

Received: 29 November 2017; Accepted: 8 January 2018; Published: 9 January 2018

Abstract: Pervious concrete is considered to be an advanced pavement material in terms of the environmental benefits arising from its basic feature—high water-permeability. This paper presents the results of experimental work that is aimed at testing technically important properties of pervious concrete prepared with three different water-to-cement ratios. The following properties of pervious concrete were tested—compressive and splitting tensile strength, unit weight at dry conditions, void content, and permeability. The mix proportions were expected to have the same volume of cement paste, and, to obtain the same 20% void content for all of the samples. The results show that changes of water-to-cement ratio from 0.35 to 0.25 caused only slight differences in strength characteristics. Arising tendency was found in the case of compressive strength and a decreasing tendency in the case of splitting tensile strength. The hydraulic conductivity ranged from 10.2 mm/s to 7.5 mm/s. The values of both the unit weight and void content were also analysed to compare the theoretical (calculated) values and real experiment results. A fairly good agreement was reached in the case of mixtures with 0.35 and 0.30 water-to-cement ratios, while minor differences were found in the case of 0.25 ratio. Finally, a very tight correlation was found between void content, hydraulic conductivity, and compressive strength.

Keywords: pervious concrete; water to cement ratio; strength; hydraulic conductivity; void content

---

1. Introduction

People change the natural environment when they build buildings and roads. One of the most notable changes is connected with the construction of impervious areas in places that were originally permeable. Impervious areas prevent water from infiltrating the soil underneath. Examples of impervious areas include rooftops, parking lots, and roadways.

The environment is adversely affected by the integration of impermeable areas into the surface—this fact causes disruption of the natural water cycle. It consequently causes the blocking of the natural process of water infiltration through the soil—thus, in the case of storm events and snowmelts, the water runoff from the impervious surfaces are very fast. There are three main aspects of this runoff, as given in [1]: "(1) a decrease in groundwater recharge due to lack of infiltration, (2) alteration in the natural flow patterns of a drainage basin, and (3) transportation of contaminants, deposited on impervious surfaces, to receiving water bodies". This is the way how the interruption of both surface and subsurface water quantity and quality is affected [1].

With development of new urban areas, there is a great challenge in finding new ways to manage storm-water runoff. Among others, porous pavements are presented as an alternative method for storm-water control. Types of porous pavements include porous asphalt, pervious cement concrete, concrete paving-blocks, gravel paving systems, and grass paving systems, among others. According to [2], the way how the pervious pavements work lies in reduction of runoff volume by allowing for water to pass through them, be stored, and subsequently be released into the ground.

Focusing on the pervious concrete, it can be described as a material consisting of open-graded coarse aggregate, Portland cement, water, and admixtures. The basic arrangement of composition contains mainly characterization of the aggregate: size of approximately 8 mm; sand is neglected to leave the space between grains empty. A representative pervious concrete has 15% to 25% of void space [2,3].

The consistency of pervious concrete mixtures is characterized by little to no slump [4]. This inherent property is due to the low cement-paste content, allowing the creation of only a thin film coating the aggregate [5]. High-viscosity paste is needed to coat the aggregates, while resisting the drain-down of the paste. The mitigation of drain-down is pertinent for the matrix porosity to be maintained across the width of the concrete section. The structure relies on the stone-to-stone contact achieved through compaction, which allows for the cement-paste-coated aggregates to bond with one another. To achieve the proper void structure, it is recommended to use the appropriate cement paste, which should possess a low water-to-cement (w/c) ratio of about 0.20 to 0.25, in addition to superplasticizer and adequate mixing [6]. The experiment given in [7] shows a decrease in compressive strength with a decrease in w/c ratio, which is unlike the conventional dense-concrete behaviour. Another experiment shows a w/c ratio that ranges from 0.2 to 0.4. They report that pervious concrete mixtures with w/c ratios under 0.3 require water-reducing admixtures while those with w/c above 0.3 can be mixed without plasticizer [5,7]. In general, w/c in the range of 0.27–0.34 can be assumed as the most common and wide range applicable for pervious concrete mixtures [3].

Another options of pervious concrete involve the improvement of water quality in ground-water recharge [8]. Due to the storm-water runoff infiltration into the ground, the sediment is filtered and contaminants do not pass into the groundwater. Similarly, due to water infiltration through the concrete layer, pervious concrete parking lots can serve as recharge basins. Other benefits of pervious concrete are following: better road safety because of increased skid resistance, road sound dampening, and a reduction of the "heat island" effect [2,9,10].

This article presents a study that has been conducted to confirm the applicability of the w/c ratio in the range of 0.35–0.25 for locally available concrete components. The influence of w/c ratio on the key properties of pervious concrete was evaluated in terms of the following properties: compressive and splitting tensile strength, unit weight, void content, and hydraulic conductivity. The void content was assumed as constant, and, together with unit weight, was controlled for comparing the theoretical (calculated) values and real experiment results. The results were also analysed in terms of the dependence of the key properties. The specifications of correlation are given as well.

2. Materials and Methods

The experiment focused on the utilization of locally available materials (Eastern Slovakia), as follows:

- cement CEM II/A-S 42.5 R complying [11], and
- natural aggregate (river gravel type, pebbles of limestone, quartz and various metamorphits) complying [12].

To create an open structure of concrete, single-size coarse aggregate of fraction 4/8 was used. To maintain the desirable performance of pervious concrete, 7% of sand was added to the aggregate mixture. The properties of aggregate are given in Table 1. The sieve analysis of aggregate is given in Figure 1.

Table 1. Properties of aggregate used in experimental work.

Properties	Aggregate Fraction	
	0/4	4/8
Density (kg/m^3)	2600	2580
Dry rodded unit weight (kg/m^3)	1830	1580
Water absorption (%)	1.2	1.4
Void content (%)	30	39

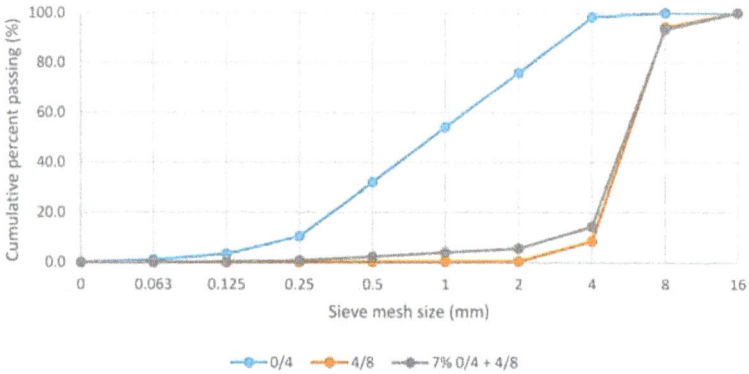

Figure 1. Sieve analysis of aggregate used in experiment.

The proportions of materials were calculated by the method given in [13]. Here, the void content is one of the input parameters. In the experiment, three recipes of pervious concrete mixtures were designed, where the void content was assumed as constant. The value was selected using the following considerations: according to [14], void content for pervious concrete should be 18–23%; according to [3], it should range between 15% and 25%. Assuming these boundaries, 20% void content was chosen for the calculation of mix proportions; hence, the volume of the aggregate (0.575 m^3) and the volume of cement paste (0.225 m^3) were constant for all of the mixtures. Mixtures differed from each other only in terms of w/c ratio (0.35, 0.30, and 0.25) and of plasticizer dosage to control the workability. Final proportions of mixtures for 1 m^3 of fresh pervious concrete and the theoretical (calculated) unit weight are given in Table 2.

Table 2. Proportions of pervious concrete mixtures per 1 m^3.

Material	Units	Mixture		
		W35	**W30**	**W25**
Cement		327	354	384
Water		115	106	96
Aggregate 0/4	(kg/m^3)	96	96	96
Aggregate 4/8		1375	1375	1375
Plasticizer		0.3	1.4	4.2
w/c	(–)	0.35	0.30	0.25
Calculated unit weight	(kg/m^3)	1913	1931	1951
Calculated void content	(%)	20	20	20

Samples of pervious concrete were mixed in the forced-action concrete mixer. The first step was to mix the aggregate for about 15 s. Then, cement was added and mixed for another 15 s. Finally, the water with superplasticizer was added and mixed thoroughly for 2 min, by a 3-min rest, and followed by a 1-min final mixing. Cylindrical specimens of 100 mm in diameter and 200 mm in height were prepared in three layers, through constant compacting (15 hits for each layer). This compaction process was chosen to determine if it is appropriate to achieve a 20% void content. Samples were demoulded after 24 h and were then cured under standard moisture and temperature conditions (water curing and 20 ± 3 °C) until the corresponding testing time. The strength properties were tested after 2 and 28 days of curing. Pace rate during compressive strength test was set to 0.6 MPa/s and 0.06 MPa/s for splitting tensile strength respectively. In addition, the compressive strength, as well as dry unit weight, void content and hydraulic conductivity were tested after 120 days of curing. The void content, together with unit weight were controlled by comparison of the theoretical (calculated) values and real experimental results. The results were also analysed by the dependence of individual properties and the specification of correlation.

The unit weight of pervious concrete specimens was tested under oven-dry conditions. Samples of pervious concrete were dried in the oven until they achieved a constant mass. That means the difference between mass of specific sample after 24 h of drying is less than 0.1%. The void content of pervious concrete specimens was tested and final values were calculated using following equation:

$$V_r = \left[1 - \frac{w_2 - w_1}{\rho_w V} \right] 100$$

where:

V_r—void content (%)
w_2—oven dried mass of sample (kg)
w_1—mass of sample submerged in water (kg)
ρ_w—density of water (kg/m^3)
V—volume of sample (m^3)

The hydraulic conductivity of pervious concrete specimens was tested using the falling-head permeability test. For this purpose, our own apparatus was constructed, as illustrated in Figure 2. The initial water level in the stand-pipe was 350 mm and the final water level was 50 mm. The time needed for water level to fall from 350 mm to 50 mm was recorded. Values of hydraulic conductivity were calculated and were expressed using the following equation:

$$k = \frac{aL}{At} \ln \frac{h_0}{h_1}$$

where:

k—hydraulic conductivity (mm/s)
a—cross section area of stand pipe (mm^2)
L—length of specimen (mm)
A—cross section area of specimen (mm^2)
t—elapsed time (s)
h_0—water head height at the beginning of the test (mm)
h_t—water head height at the end of the test (mm)

Initial water level in stand pipe was 350 mm and final water level was 50 mm. Time needed to fall water level from 350 mm to 50 mm was recorded.

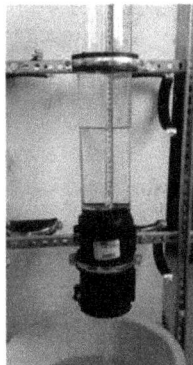

Figure 2. The apparatus and arrangement for testing the hydraulic conductivity of pervious concrete sample.

3. Results and Discussion

In this chapter, the results in terms of compressive strength, splitting tensile strength, unit weight, void content, and hydraulic conductivity are given. The appropriate discussion is attached to the results where unexpected behaviour was observed. All of the results are given as an average of four samples, while the standard deviation is given in the parentheses. The relationships between the key features of pervious concrete are given in the end of this chapter.

During the mixing, the workability of fresh mixture was visually controlled. Mixture W35 ($w/c = 0.35$) showed a slightly runny character immediately after mixing; however, after a few minutes, the excessive water was soaked by the aggregate. On the other hand, mixture W25 ($w/c = 0.25$) showed a very stiff nature and despite the higher dosage of plasticizer (lignosulphonate base), it was difficult to process it.

3.1. Compressive and Splitting Tensile Strength

The compressive and splitting tensile strength of concrete samples were tested. Compressive strength was tested after 2, 28, and 120 days of curing, while the splitting tensile strength was tested after 2 and 28 days of curing (Table 3).

Table 3. Results of compressive (f_c) and splitting tensile strength (f_t) including standard deviation.

Sample	f_c after 2 Days (MPa)	f_t after 2 Days	f_c after 28 Days (MPa)	f_t after 28 Days	f_c after 120 Days (MPa)
W35	9.5 (0.9)	1.3 (0.10)	14.5 (0.3)	2.0 (0.21)	20.0 (4.3)
W30	9.0 (0.6)	1.5 (0.21)	16.0 (1.2)	1.7 (0.53)	19.0 (1.9)
W25	10.5 (1.7)	1.4 (0.23)	17.5 (2.8)	1.6 (0.26)	19.5 (3.7)

The highest compressive strength after two days of curing was achieved by the W25 mixture (10.5 MPa). The highest compressive strength after 28 days of curing was also achieved by the W25 mixture (17.5 MPa). After 120 days of curing, the highest compressive strength was achieved by the W35 mixture (20.0 MPa). The differences between the compressive strength values of mixtures are very slight and can probably be interpreted as negligible from the point of view of w/c ratio range.

The highest splitting tensile strength after two days of curing was achieved by the W30 mixture (1.5 MPa). The highest splitting tensile strength after 28 days of curing was achieved by the W35 mixture (2.0 MPa). Since this mixture has the highest w/c (0.35), it probably can be explained by better workability, and thus better bond between the aggregate and the cement paste. In the case of the W30 and W25 mixtures, stiffer cement paste showed a weaker adhesion to aggregate grains, which resulted in lower splitting tensile strength. Similar behaviour is observed and discussed in [15,16]. However, due to the very slight difference in splitting tensile strength between mixtures (difference was smaller than measurement accuracy), it can be assumed that the present variation in the w/c ratio does not have a significant influence on the splitting tensile strength. Testing the splitting tensile strength enables the study of the nature of failure, since it is clearly visible. The failure crack came through most of the aggregate grains; this indicates the good performance of cement stone. Therefore, the strength of the aggregate seems to be a limiting factor here, and further increasing of the cement-paste strength by lowering the w/c ratio would not necessarily lead to a better strength of concrete, as also shown in [16].

3.2. Density and Void Content

The results of unit weight at dry conditions and void content are given in Table 4. Since these properties directly affect the permeability, it is first evaluated whether they meet the theoretical values. Mixtures W35 and W30 are close to the calculated unit weight (given in Table 2), while mixture W25 achieved slightly lower unit weight than expected. According to the calculations and theoretical assumptions, the unit weight should increase as the w/c ratio decreases. This behaviour is not observed in our experiment. It can be explained by application of the same way of compaction for all of the samples, while the stiffer W30 and W25 mixtures require more compaction effort to reach the target unit weight. The void content of mixtures did not exactly achieve the calculated values (20%); the values were higher—23% for mixtures W35 and W30 and 26% for the W25 mixture. The void content theoretically should not change with w/c ratio variations because it comes from the mixture design method, and because all the mixtures had the same volume of cement paste and aggregate. The same explanation of different behaviour than expected can be stated here as above. However, it only describes consistency between theoretical calculation and real experiment results. What is more important is that the values of void content are practically the same, as expected.

Table 4. Results of unit weight, void content, and hydraulic conductivity, including standard deviation.

Sample	ρ_v (Dry) (kg/m^3)	V_r (%)	k (mm/s)
W35	1930 (56)	23 (3.9)	8.6 (6.3)
W30	1930 (36)	23 (1.8)	7.5 (1.2)
W25	1890 (42)	26 (2.6)	10.2 (2.1)

3.3. Hydraulic Conductivity

The hydraulic conductivity is the basic parameter to define the permeability of pervious concrete. The results of hydraulic conductivity are given in Table 4. Values are comparable with those presented in studies like [17,18]. The highest hydraulic conductivity was achieved by the W25 mixture (10.2 mm/s). As in the case of void content, the hydraulic conductivity should be the same for all of the mixtures and should not vary with changes in w/c ratio, since it is mainly the function of interconnected void content, which has been assumed in the calculation at 20% for all of the mixtures. In practice, the results of hydraulic conductivity in the experiment on average corresponds with those of void content given in Table 4—higher is the void content, higher is the hydraulic conductivity. Similarly, it is presented in [17] that the mixture with the highest hydraulic conductivity supports the conclusion that the low w/c ratio would have led to decreased workability and lower density. This lower density results in a greater amount of pore space available for water to pass through, thus increase the hydraulic conductivity.

A wide range of hydraulic conductivity results was achieved with the W35 mixture (2.4–15 mm/s), which is represented by high standard deviation. This is probably a consequence of higher w/c ratio (0.35)—it makes the cement paste too runny and affects the void structure, and thus also the hydraulic conductivity in an inappropriate way (the cement paste settles down and closes the open void structure). It could be hard to repeat the same results with the same mixture.

3.4. Interactions between Pervious Concrete Properties

The dependence between void content, hydraulic conductivity, and compressive strength is illustrated in the Figure 3, while the exponential regression model was applied. The experiment shows very tight dependence between void content and hydraulic conductivity ($R^2 = 0.73$), as well as between void content and compressive strength ($R^2 = 0.72$).

The dependence between unit weight, compressive strength, and void content is illustrated in the Figure 4, while the linear regression model was applied. The experiment shows very tight dependence between unit weight and compressive strength ($R^2 = 0.74$), as well as between unit weight and void content ($R^2 = 0.94$).

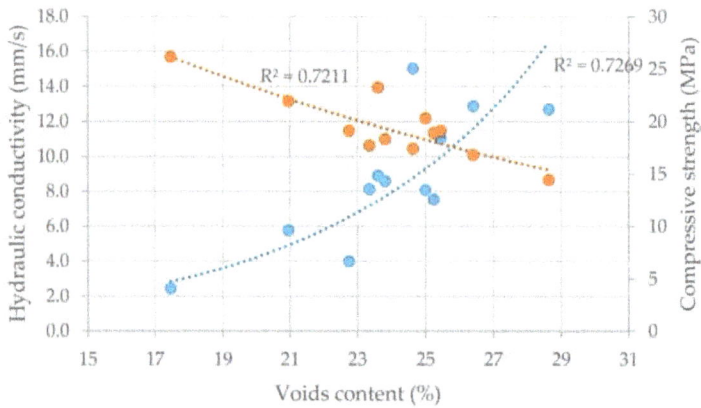

Figure 3. Interaction between hydraulic conductivity, void content, and the compressive strength of pervious concrete samples.

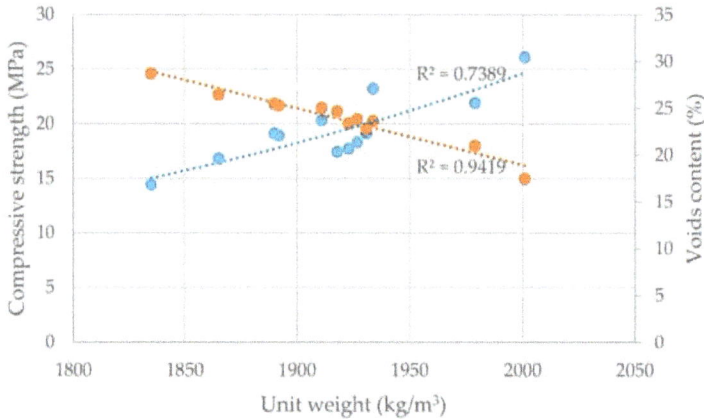

Figure 4. Interaction between compressive strength, unit weight and void content of pervious concrete samples.

4. Conclusions

The experiment shows results pertaining to the key properties of pervious concrete made of locally available materials. As such, the test results could be useful for others working in the same area. The influence of w/c ratio in the range 0.35 to 0.25 on pervious concrete properties was tested. The findings should be weighted from the point of view of specific approach to design of mix proportions that were used in this experiment—the constant volume of the aggregate and the cement paste for all tested mixtures. Changes in the w/c ratio were made not only by changing the amount of water to the same amount of cement, but the amount of cement was corrected accordingly to keep the same volume of cement paste. The following conclusions can be formulated:

- Decrease in w/c ratio caused the fresh pervious concrete to be stiffer, and thus more difficult to process—despite the use of higher volume of plasticizer. This probably resulted in unexpectedly lower unit weight, higher void content, and higher hydraulic conductivity of the W25 mixture.

- The tested range of w/c ratio caused only very slight differences in strength characteristics of pervious concrete, though practically sufficient values were achieved: 14.5–17.5 MPa of compressive strength and 1.6–2.0 MPa of splitting tensile strength.
- The strength of the aggregate seems to be a limiting factor in the further strengthening of pervious concrete. This opinion is based on the crack formation that occurs at yield strength of samples.
- When applying 0.35–0.25 w/c ratio, hydraulic conductivity values ranging from 8.6 to 10.2 mm/s were achieved.
- A good correlation was found for the key properties of pervious concrete, namely unit weight, compressive strength, void content, and hydraulic conductivity. The regression analysis shows high R^2 values (0.72–0.94).
- Optimization of the kind and dose of chemical admixtures is necessary for the production of pervious concrete with very low w/c ratio.

The results of strength characteristics found in the presented experiment are promising and open up opportunities for future experimental work focusing on various locally available materials. The achieved values of hydraulic conductivity can be usable in storm-water management of urban areas when applying the pervious concrete for pavements, thus bringing environmental benefit to the lives of people who reside in cities.

Acknowledgments: This work was supported by the project ITMS Center of excellent integrated research of progressive building constructions, materials and technologies (grant number 26220120037) and by the project VEGA (grant number 1/0524/18).

Author Contributions: Marek Kováč conceived, designed and performed the experiment, analyzed data and wrote the paper; Alena Sičáková wrote the paper. Both authors read and approved the final manuscript.

Conflicts of Interest: The authors declare no conflict of interest.

References

1. Brattebo, B.; Booth, D. Long-Term Stormwater Quantity and Quality Performance of Permeable Pavement Systems. *Water Res.* **2003**, *37*, 4369–4376. [CrossRef]
2. Brown, D. Pervious Concrete Pavement: A Win-Win System. *Concr. Technol. Today* **2003**, *24*, 1–3.
3. Tennis, P.D.; Leming, M.L.; Akers, D.J. *Pervious Concrete Pavements*; Portland Cement Association: Skokie, IL, USA, 2004.
4. Haselbach, L.; Freeman, R. Vertical Porosity Distribution in Pervious Concrete Pavement. *ACI Mater. J.* **2006**, *103*, 452–458.
5. Wang, K.; Schaefer, V.R.; Kevern, J.T.; Suleiman, M.T. *Development of Mix Proportion for Functional and Durable Pervious Concrete*; NRMCA Concrete Technology Forum: Nashville, TN, USA, 2006.
6. Chindaprasirt, P.; Hatanaka, S.; Chareerat, T.; Mishima, N.; Yuasa, Y. Cement Paste Characteristics and Porous Concrete Properties. *Constr. Build. Mater.* **2008**, *22*, 894–901. [CrossRef]
7. Joshi, T.; Dave, U. Evaluation of Strength, Permeability and Void Ratio of Pervious Concrete with Changing w/c Ratio and Aggregate Size. *Int. J. Civ. Eng. Technol.* **2016**, *7*, 276–284.
8. Legret, M.; Colandini, V.; Le Marc, C. Effects of a Porous Pavement with Reservoir Structure on the Quality of Runoff Water and Soil. *Sci. Total Environ.* **1996**, *190*, 335–340. [CrossRef]
9. Yang, J.; Jiang, G. Experimental Study on Properties of Pervious Concrete Pavement Materials. *Cem. Concr. Res.* **2003**, *33*, 381–386. [CrossRef]
10. The United States Environmental Protection Agency (US EPA). *Preliminary Data Summary of Urban Storm Water Best Management Practices*; The United States Environmental Protection Agency: Washington, DC, USA, 1999.
11. Slovak Office of Standards, Metrology and Testing. *STN EN 197-1 Cement. Part 1: Composition, Specifications and Conformity Criteria for Common Cements*; Slovak Office of Standards, Metrology and Testing: Bratislava, Slovakia, 2011.
12. Slovak Office of Standards, Metrology and Testing. *STN EN 12620: Aggregates for Concrete*; Slovak Office of Standards, Metrology and Testing: Bratislava, Slovakia, 2008.

13. Nguyen, D.H.; Sebaibi, N.; Boutouil, M.; Leleyter, L.; Baurd, F. A Modified Method for the Design of Pervious Concrete Mix. *Constr. Build. Mater.* **2014**, *73*, 271–282. [CrossRef]
14. Slovak Office of Standards, Metrology and Testing. *STN 73 6124-2: Road Construction. Part 2: Concrete Drainage Layers*; Slovak Office of Standards, Metrology and Testing: Bratislava, Slovakia, 2013.
15. Neamitha, M.; Supraja, T.M. Influence of Water Cement Ratio and the Size of Aggregate on The Properties Of Pervious Concrete. *Int. Ref. J. Eng. Sci.* **2017**, *6*, 9–16.
16. Chopra, M.; Wanielista, J.; Mulligan, A.M. *Compressive Strength of Pervious Concrete Pavements*; Storm Water Managenent Academy, University of Central Florida: Orlando, FL, USA, 2007.
17. McCain, G.N.; Dewoolkar, M.M. Porous Concrete Pavements: Mechanical and Hydraulic Properties. Available online: http://www.uvm.edu/~transctr/publications/TRB_2010/10-2228.pdf (accessed on 29 November 2017).
18. Batezini, R.; Balbo, J.T. Study on the hydraulic conductivity by constant and falling head methods for pervious concrete. *Revista IBRACON de Estruturas e Materiais* **2015**, *8*. [CrossRef]

MDPI

St. Alban-Anlage 66

4052 Basel

Switzerland

Tel. +41 61 683 77 34

Fax +41 61 302 89 18

www.mdpi.com

Environments Editorial Office

E-mail: environments@mdpi.com

www.mdpi.com/journal/environments